Emerging Technologies
in Plastics Recycling

ACS SYMPOSIUM SERIES **513**

Emerging Technologies in Plastics Recycling

Gerald D. Andrews, EDITOR

E. I. du Pont de Nemours and Company

Pallatheri M. Subramanian, EDITOR

E. I. du Pont de Nemours and Company

Developed from a symposium sponsored
by the Division of Polymer Chemistry, Inc.,
of the American Chemical Society
at the Polymer Technology Conference,
Philadelphia, Pennsylvania,
June 3–5, 1991

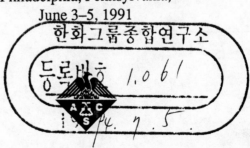

American Chemical Society, Washington, DC 1992

Library of Congress Cataloging-in-Publication Data

Emerging technologies in plastics recycling / Gerald D. Andrews, editor, Pallatheri M. Subramanian, editor.

p.   cm.—(ACS symposium series, ISSN 0097–6156; 513)

"Developed from a symposium sponsored by the Division of Polymer Chemistry, Inc., of the American Chemical Society at the Polymer Technology Conference, Philadelphia, Pennsylvania, June 3–5, 1991."

Includes bibliographical references and indexes.

ISBN 0–8412–2499–4

1. Plastics—Recycling—Congresses.

I. Andrews, Gerald D., 1948–   . II. Subramanian, Pallatheri M., 1931–   . III. American Chemical Society. Division of Polymer Chemistry. IV. Series.

TP1122.E44   1992
668.4–dc20   92–27130
                                                             CIP

The paper used in this publication meets the minimum requirements of American National Standard for Information Sciences—Permanence of Paper for Printed Library Materials, ANSI Z39.48–1984. ∞

Foreword

THE ACS SYMPOSIUM SERIES was first published in 1974 to provide a mechanism for publishing symposia quickly in book form. The purpose of this series is to publish comprehensive books developed from symposia, which are usually "snapshots in time" of the current research being done on a topic, plus some review material on the topic. For this reason, it is necessary that the papers be published as quickly as possible.

Before a symposium-based book is put under contract, the proposed table of contents is reviewed for appropriateness to the topic and for comprehensiveness of the collection. Some papers are excluded at this point, and others are added to round out the scope of the volume. In addition, a draft of each paper is peer-reviewed prior to final acceptance or rejection. This anonymous review process is supervised by the organizer(s) of the symposium, who become the editor(s) of the book. The authors then revise their papers according the the recommendations of both the reviewers and the editors, prepare camera-ready copy, and submit the final papers to the editors, who check that all necessary revisions have been made.

As a rule, only original research papers and original review papers are included in the volumes. Verbatim reproductions of previously published papers are not accepted.

M. Joan Comstock
Series Editor

Contents

INDEXES

Preface

CONCERNS FOR THE ENVIRONMENT and the limited resources of the Earth have finally awakened many of its citizens. They now recognize the need for more efficient and conservative use of resources. Recycling can be an effective way to conserve global material resources and minimize disposals in landfills. We currently recycle less than 500 million kg of the worldwide production of about 60 billion kg of plastics. Significant amounts of industrial plastic waste are recycled, but this recycling is not very visible, and meanwhile, consumer waste from packaging, automobiles, and durable goods is filling up landfills.

In 1989, approximately 150 million kg of postconsumer plastics was recycled in the United States versus 140 million kg in 1988. The vast majority of this was poly(ethylene terephthalate) from soda bottles and X-ray film, and polyethylene detergent and milk bottles. The quality of these recycled plastics varies greatly as a result of impurities such as paper, adhesives, residues, other polymers, and metals. The polymers themselves differ significantly in molecular weight, branching, and additives. Recycling of plastics from discarded automobiles and other durable goods is in its infancy. Major manufacturers of plastics for durable goods are beginning to address recycling issues.

The accelerating pace of polymer recycling technology development shows the need for a book of this type. We hope that this book stimulates scientific research in polymer recycling by summarizing the scope of the problem, highlighting current areas of activity, and demonstrating that science and technology have critical roles in helping decide what problems are worth solving and then providing the technical parts of the solutions. Development of the science and technology represented here is necessary to support the increasing public demands for recycling waste plastics in all forms.

No symposium or book of this sort could start to cover the field of polymer recycling in a comprehensive way. We therefore tried to select a range of topics that represent activities in the field. The first section is a selection of general chapters covering the background of recycling and outlining the major areas of current activities. The remaining sections cover topics from basic science development to technology application in specific polymer systems.

Polymer recycling is an emerging technology that will continue to grow. Research on molecular design of more recyclable polymers and additives (such as colors and pigments) and a focus on recyclability of products after use will enable development of materials that are more environmentally benign. Economical recycling of durable goods will require advances in the technologies for sorting, identifying, and purifying mixed materials. The infrastructure for efficient collection and recycling of plastic must be established. Much work remains to be done to make plastic recycling a significant contributor to conservation of our limited natural resources, and research on recycling is the foundation of this work.

Acknowledgments

We thank all the authors for the time and effort they spent in writing the chapters for the book. We also thank the speakers at the symposium; they contributed to a very successful meeting. Marge Maney helped organize the symposium program, and Patricia Veit helped us manage the paperwork involved in the production of the book.

GERALD D. ANDREWS
Du Pont Polymers
Wilmington, DE 19880–0323

PALLATHERI M. SUBRAMANIAN
Du Pont Polymers
Wilmington, DE 19880–0323

April 2, 1992

Chapter 1

Recycling Plastics from Municipal Solid Waste
An Overview

Wayne Pearson

Plastics Recycling Foundation, P.O. Box 189, Kennett Square, PA 19348

Recycling means that we are going to try to sell our garbage, or other disposables, to someone. The problem is that garbage has very little value. In fact, if it were extremely valuable, it would be stolen. It is a fact that no one wants to buy garbage.

Consequently, a lot of work must be done to the garbage pile to turn it into viable products. This work includes processing the garbage to quality raw materials that can be manufactured into products that can be sold to buyers who will buy it again, and again, and again.

Recycling requires four elements:

- collection
- sorting of raw materials
- reclaiming the raw materials to make a product
- markets and paying customers for the product

The Nation is producing 180 tons of municipal solid waste a year.

Based on the EPA study, about 29 billion pounds of plastic entered the waste stream in 1988. This represented about 56% of what was produced that year. The lag is what would be expected, considering the different life cycles of the different uses.

Because the life cycle of packaging is so very short, we can assume that virtually all of the production ends in the waste stream within one year, whereas the more durable uses would take longer to reach the waste stream.

The EPA claims that eight percent of the 180 million tons of MSW or about 29 billion pounds of plastics in municipal solid waste accounts for about 20 percent of the volume (Figures 1 and 2).

0097–6156/92/0513–0001$06.00/0
© 1992 American Chemical Society

Packaging, the element in the waste that is getting the most national attention and the highest concern, represents about 30% (Figures 3 and 4).

The plastics share of packaging by weight is 13 percent (Figure 5). Packaging is dominated by paper and glass (Figure 6).

The following recyclable items are reasonably viable in terms of technology and economics today:

- newspapers
- aluminum and steel cans
- glass bottles and jars
- plastic beverage containers

It's interesting to note that on a volume basis the plastic beverage container is equal to about 1/3 of the volume of things that are recyclable today, namely: newspapers, non-plastic beverage containers and plastic beverage containers (Figure 7).

Moreover, it is technically and economically viable to collect and sort plastic beverage bottles to reclaim the polymers contained therein which include polyethylene terephthalate (PET), high density polyethylene (HDPE) and polyvinyl chloride (PVC). The markets for these reclaimed polymers exist in far greater size than all the polymer used in the manufacture of beverage containers. Including plastics beverage containers with non-plastic beverage containers (glass, aluminum and steel) along with newspapers increases the volume that can be recycled by the home owner by about 50 percent, and it can be highly profitable because plastics beverage containers are second only to aluminum in value.

This technology to collect and sort these commodities is available to hundreds of communities in the Nation who are redeveloping and implementing plans to collect and include them in their recycling programs.

The infrastructure for reclaiming, cleaning up and producing products to sell is growing fast. There are a number of major companies involved in the business already, and the list is growing (Table 1).

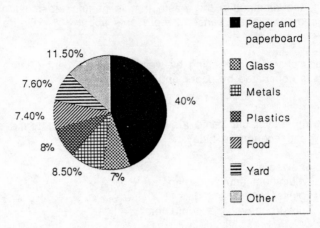

Figure 1. Materials Generated in MSW by Weight (180 million tons), 1988.

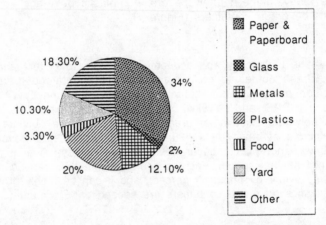

Figure 2. Materials in MSW by Volume (Cubic Yards), 1988.

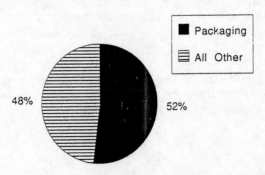

Figure 3. Percent of Plastics in MSW by Use (%).

TABLE 1. Major U. S. and Canadian Scrap
Plastic Bottle Reclaimers (1991)

Company	Location	Estimated Annual Capacity in Million Lbs.
Clean Tech	Dundee, MI	12
Day Products	Bridgeport, NJ	40
Eaglebrook Plastics	Chicago, IL	35
Graham Recycling	York, PA	20
Johnson Controls	Novi, MI	20
M.A. Industries	Peachtree City, GA	20
Midwest Plastics	Edgerton, WI	10
Orion Pacific	Odessa, TX	7
Partek	Vancouver, WA	5
Pelo Plastique	Berthierville, PQ	15
Plastic Recycling Alliance	Philadelphia, PA	40
Plastic Recycling Alliance	Chicago, IL	40
Quantum Chemical Co.	Heath, OH	40
Star Plastics	Albany, NY	32
St. Jude Polymer	Frackville, PA	25
Union Carbide	Piscataway, NJ	50
United Resource	Findlay, OH	8
United Resource Recovery	Johnsonville, SC	175

Source: Resource Recycling, 1991

In addition, the National Polystyrene Corporation has several polystyrene plants and more in the planning. There are also a number of manufacturers of plastic profiles for park and highway equipment using commingled plastic from the waste stream.

Some driving forces are accelerating and expanding the demand for these recycled materials, particularly the activities of packagers who wish to use recycled plastic to achieve a more environmentally attractive package.

In a word, the Nation is well positioned to recycle newspapers and beverage containers, including plastic beverage containers.

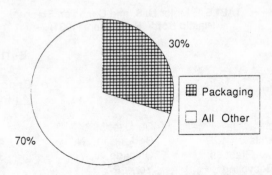

Figure 4.   Packaging Share of Municipal Solid Waste.

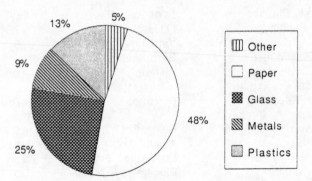

Figure 5.   Plastics Share of Packaging.

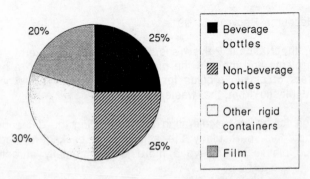

Figure 6.   Plastics Packaging by Volume.

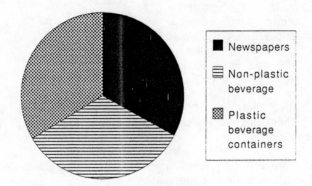

Figure 7.   Today's Recyclables by Volume.

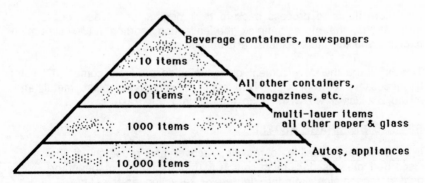

Figure 8.   Trash Pile-Number of Items.

However, to fully understand the complexity of processing trash to useful products beyond these traditional items, it is helpful to gain a perspective. The trash pile can be viewed as a pyramid as shown below (Figure 8). At the top are ten items we commonly recycle, namely:

- newspapers
- aluminum beverage cans
- steel beverage cans
- tin cans
- clear glass bottles
- brown glass bottles
- amber glass bottles
- green glass bottles
- PET soft drink bottles
- HDPE milk, water and juice bottles

At the next level there are 100 items which include: colored plastic detergent bottles, pharmaceutical bottles, other papers, magazines, and bi-metal containers.

Below that there are 1,00 items which include other glass, ceramics, multi-layer materials (foil, paper, etc.).

At the next level there may be 10,000 items including household clothing, industry, etc.

The more items to process back to their original or "closed loop" applications, the more complex the collection, sorting and reclamation become and the more expensive it becomes.

This is where the development of commingled products comes in. It may be necessary to work with mixed paper, mixed glass, mixed metal, and mixed plastics if we wish to go deeper into our trash.

BEYOND BEVERAGE BOTTLES

It is technically feasible to recycle, recover and reuse all of the discarded plastics packaging, but with the exception of beverage bottles, economics limit the degree of recycling at this time. Communities are not enthusiastic about collecting materials if their alternate methods of garbage disposal cost less. Therefore, it is necessary to consider the cost to the community as well as the cost to the reclaimer to get the full cost of the recycled material. The fact that the recycled material must compete with virgin resin in quality and price must be taken into account when considering markets.

The cost to the community includes the following:

- cost to collect recyclables
- cost to sort recyclables

From these two costs the community is entitled to subtract:

- the cost of collecting trash that is avoided (by diverting it to recyclables)
- the cost of disposal avoided (e.g. tipping fee at landfill)

The sum of these four items will be the cost to the community. The community can recover some or all of that cost by selling the recovered recyclables. The price that they need to cover the total cost is the theoretical price the reclaimer should pay for a raw material.

The reclaimer will have the following costs:

- raw material (equals community cost)
- reclaiming cost (e.g. cost to convert the raw material to product
- marketing expenses and overheads
- profit and return on investment

The sum of all these costs, including dividends needed to pay stockholders, becomes the minimum price required to make an economically attractive venture for manufacturing a product derived from feedstock obtained from processing trash.

The next question becomes: Is the price higher or lower than the price of virgin material? The processing technologies today are good enough to manufacture a product essentially equal in quality to virgin. However, in a normal free market situation a manufacturer will insist on receiving a discount to use what is deemed to be inferior quality material.

If we take a look at the economic viability of plastics packaging recycling considering the virgin prices based on oil at nominally $20/barrel (Figure 9) we see clearly that beverage containers are economically viable with current cost to landfill which is between $50 and $100 ton. However, it is also clear that beyond beverage bottles it is not economically attractive at today's price of oil, alternative disposal costs and current recycling technology.

If, however, the price of oil were to rise to $35 per barrel (Figure 10), then the cost associated with manufacturing virgin polymers from high priced oil-related feedstocks would increase substantially. Therefore, even if recycled material were discounted, versus virgin prices, the price would be high enough to make it economic to collect, sort, and reclaim many plastic components from the waste stream.

This suggests that there is a "future value" to recycling that should be given consideration from the point of view of "stock piling" material going into the trash stream to be utilized in a future time when the price of oil might be higher than $20/barrel.

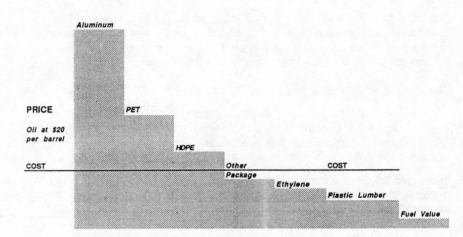

Figure 9. Viability of Plastics Packaging Recycling vs. Today's Virgin Prices Based on Oil at $20/Barrel.

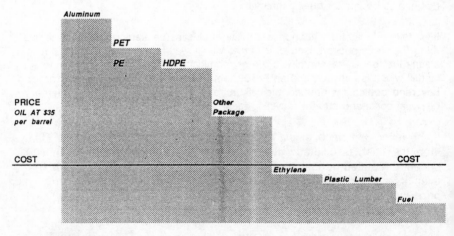

Figure 10. Viability of Plastics Package Recycling vs. Virgin Prices Based on Oil at $35/barrel.

Another issue that surfaces here is that if the public decides to purchase recycled material, despite quality deficiencies, then the price of recycled material need not be discounted versus virgin. In fact, the price of recycled material could exceed the price of virgin. That would change economics substantially, as shown with non-beverage bottles (Figure 11).

This new marketing factor can be labelled "green". Simply stated, it means that a product that can trace its pedigree to the post-consumer garbage pile will be preferentially purchased versus one whose pedigree is traced to natural gas or oil.

RESEARCH PROGRAMS

To go beyond beverage bottles is needed to develop next generation technology. We must reduce the cost of collection and sorting and upgrade the quality of the generics recovered through improved processing technologies so as to increase the value.

The process for deriving products with a garbage pile origin can be diagrammed as follows:

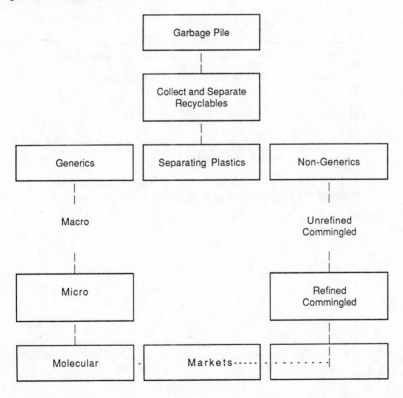

PRICE

Cents/Lb.

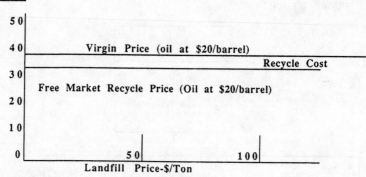

Figure 11.   Economic Viability-Plastics Package
Recycling/Non-Beverage Bottles.

Once the plastics are separated from the non-plastic recyclables, there are a number of technologies for separation of the plastics to generic components and there are technologies for dealing with commingled (unseparated) plastics.

Research programs that are underway focus on the issue of collection, sorting/reclaiming and end-use markets.

COLLECTION

In collection the issues are:

- collection of all clean plastics
- collection container
- packer trucks
- truck routing

The question is: "What will the consumer give when asked to deliver all clean plastics?" This includes all rigid containers (beverage bottles, detergent bottles, and rigid tubs) and non-rigid materials such as film. Types of trucks and truck routing are also being studied.

SORTING/RECLAIMING

In the sorting and reclaiming of generics, the research is focused on three types of separation:

- macro
- micro
- molecular

Macro-sorting is the separation at the container level. Micro-sorting is the separation of chips of containers after the containers have been shredded and chopped up. Molecular sorting involves dissolving polymers in an organic solvent. It has been demonstrated that five or more polymers can be separated for packaging from each other in a solvent system.

Some polymers will be de-polymerized (sorted at the monomer molecule level). Condensation polymers are among the more easily de-polymerized molecules, so it is not surprising to see commercial operation in place or planned to de-polymerize PET (from SODA BOTTLES) to the monomers dimethyl terephthalate and ethylene glycol.

Purification of these molecular separations can be high. Monomer derived from post-consumer PET bottles has been found acceptable by the FDA.

While molecular separation technology may not be economically viable today, it will become more economically viable as the price of oil increases, and/or as costs can be reduced for this methodology.

Industry and recycling communities are working on these same issues so that the number of high quality generic plastic items in the trash pile that will become economically viable will increase with time.

In connection with minimizing cost of collecting and sorting, technologies that simplify collection, sorting and cleaning are surfacing. Two of these technologies are:

- the molecular separation of commingled plastics
- the development of new compounds and families of products based on "refined commingled" plastics

It has been demonstrated at the Center for Plastics Recycling Research at Rutgers University, that mixed or commingled plastics from the post-consumer waste stream can be used to make a wide range of plastic products using traditional virgin plastics processing equipment.

The commingled plastic material requires no macro sorting and collecting can be simplified. Thus the cost of collection and sorting can be reduced substantially.

Once collected, the commingled plastic materials are "refined" by chopping the plastics and washing them free of paper, metal and any other non-plastic materials.

The "refined" (clean) commingled materials can then be modified by mixing in one of the polymers or compatibilizers to achieve desired end properties. The compatibilized mixture is melt-filtered as it is blended to produce a high purity pellet which can be processed on conventional plastics processing equipment.

It is desirable to segregate blow molding resin types from injection or extrusion resins which can be accomplished at the "refined" commingled plastic reclamation plant. Thus a broad range of "new" products can be designed which expand the recycling potential for post-consumer plastics.

There are a few commercial products we would define as "refined commingled". These include mixtures of reclaimed polymer with virgin polymer and compatibilizers added to achieve a specific property.

These molecular separation and refined commingled technologies do not require plastics to be sorted. Moreover, the range of plastics collected from the householder could be enlarged substantially. Bales containing a wide range of unsorted post-consumer plastics would be shipped to a plant that could separate them into generic resins at the molecular level or could manufacture new compounds from the commingled, unseparated materials.

Both of these technologies will drive the cost of collecting and sorting down and at the same time will provide some reasonable level of value for the recovered generics and/or non-generic (commingled) plastics.

END USE

End use market research programs continue to focus on and expand the market opportunities for the full range of polymers from the waste including polyethylene terephthalate (PET), high density polyethylene (HDPE), polyvinylchloride (PVC), polystyrene (PS), and polypropylene (PP), and market research on non-generic uses including the previously mentioned new family of compounds and profiles is continuing.

CONCLUSION

In conclusion, in 1985, very few people believed that plastics from the post-consumer stream were recyclable, let alone were being recycled. Today, the amount that is recycled is still in an early stage. Approximately 300 million pounds of PET bottles are being recycled, which represents about 30% of that container, and probably about 10% of the milk, water and juice bottles are being recycled. However, the technology has been researched sufficiently so that we can say confidently that the entire packaging stream is recyclable. That is to say, we have the technology for collecting, sorting, cleaning it up and finding markets for it.

The economics for recycling plastic beverage containers are very favorable making it highly desirable for communities to include them in their recycling programs right now. Should they elect to do that, they will be able to have the potential for increasing the volume they can remove from the waste stream by 50%, assuming that they are collecting newspapers and non-beverage containers.

Looking beyond plastic beverage containers, we see that the technology is viable but that there is a lot of work to do to reduce the cost of collection, sorting and reclamation, and there is also work to do on the market development. However, with time, the rising costs of virgin material, and alternative disposal costs will provide the driving forces that will increase the percent of post-consumer plastics that are recycled.

RECEIVED June 10, 1992

OVERVIEW OF RECYCLING TECHNOLOGY

Chapter 2

Environmental Compatibility of Polymers

F. Peter Boettcher

Du Pont Polymers, Experimental Station 323/325,
Wilmington, DE 19880–0323

On the list of options for the environmentally safe disposal of plastics, recycling has a preferred position because it recovers most of the inherent value. Recycling of post-consumer plastic waste as a systematic effort is still too new to judge its ultimate economic success. Eventually, recycling can only succeed if supply and demand are in balance and an economic driving force exists. It is clear, however, that this will only occur if the quality of recycled materials is very high so that they can compete effectively for already existing market needs. Proper part design and selection of plastic type can significantly enhance chances for cost-effective material recovery. If purification is needed to restore value-in-use, chemical or thermal cleavage of the plastic material with recovery of the monomers is an option. However, there will be some portions of post-consumer plastic waste which can not be recovered economically with currently available technology. In this case, incineration with energy recovery should be given preference over landfilling.

The environmental impact of discarded products based on man-made materials is receiving increasing public attention. Demands for products which are compatible with the environment have rapidly accelerated in the U.S., in Europe, and more recently in Japan. They are currently focussed on one material, i.e. plastics, despite the fact that the majority of items used in daily life today is made from materials which have been manufactured by a chemical process.

About 26 MM t of plastics were sold in the U.S. in 1989 (Fig. 1) which is somewhat less than one-third of world plastics sales. Only three types of plastics (PE, PP, PVC) made up 60% of the total volume. Estimates show that roughly 2% of our oil refinery output was needed as raw material, an almost insignificant amount compared to the consumption of gasoline and fuel oils.

In assessing the impact of materials on the environment, one needs to consider all phases of their life cycle, i.e. manufacture, use and disposal. According to a 1989 study by Franklin Associates, Ltd., a comparison of air pollution and water pollution occurring during the entire life cycle of one-way PET, glass, and aluminum soft drink containers turned out favorable for PET (Fig. 2). Only refillable glass bottles, when used at least five times, had lower environmental impact. One might expect similar or

0097–6156/92/0513–0016$06.00/0

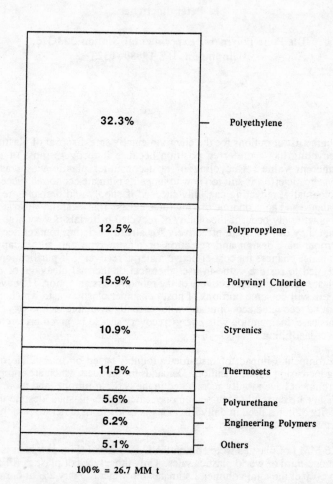

100% = 26.7 MM t

Figure 1. U. S. Plastics Sales 1989 (SOURCE: Adapted from *Modern Plastics,* January 1990)

even better results for other plastics. The study assumed 1987 recycling rates and counted all materials which were not recycled as waste. A factor often overlooked in the discussion of refillable containers is the substantial amount of polluted water generated in the washing step. This is a significantly smaller problem for plastics recycling.

In a comparison of energy consumption and volume of solid waste for manufacture, use and disposal, PET also has an advantage, again only exceeded by refillable glass bottles (Fig. 3). Because of their lower density and high mechanical properties, plastic products weigh less than their metal or glass counterparts, which can result in substantial energy savings during their life cycle. It has been estimated that fuel consumption of trucks transporting filled or empty soft drink bottles is reduced by 50% when changing from glass to PET. An example (Fig. 4) recently cited at a plastics conference at the University of Stuttgart in Germany demonstrates the large weight savings obtainable in going from glass milk bottles (31% of total) to polyethylene pouches (0.7% of total). This amounts to a 44-fold reduction of ratio of contents to packaging. Use of plastics to improve the energy efficiency of automobiles is already a classical example. For each kilogram of weight reduction, approximately 9 L of gasoline are saved during the life of a car (10 yrs, 150,000 km). In this case, there is unquestionably a large benefit in terms of reduced crude oil consumption and air pollution. Generally, the use of plastics leads to fewer polluting by-products and requires less energy than that of potential alternative materials, e.g. metal, glass, and even paper. Refined ecobalances which will take into consideration the entire environmental impact of plastics "from cradle to grave" are the subject of studies here and abroad. This information increasingly provides a rational basis for determining the environmental impact of plastics vs. other materials.

Current concerns over the impact of plastics on the environment concentrate, unfortunately, mostly on their disposal, especially disposal in landfills. There is no direct negative environmental impact from landfilled plastics because their high stability essentially guarantees that there is no interaction with the environment. However, plastics currently constitute about 18% by volume of all landfilled materials and this amount is likely to grow in the next 10 years as output grows and the amount of discarded plastics gets into equilibrium with the amount produced. It is this fact combined with the rapidly declining number of landfills which has led to the intense interest in the disposal of plastics. Just as with other materials, recycling in many cases appears to be an effective method to cope with this problem. Even if all plastics were only used twice, i.e. recycled once, the amount of discarded plastics would be reduced by 50%.

Iron, aluminum, and glass have already reached high recycling rates (Fig. 5). Here, the residual economic value of the used material provides as much driving force for recycling as the desire to reduce negative environmental impact. That this can also be a consideration for plastics is shown by the fact that recycling of PET soft drink bottles started very soon after their introduction. Today in the U.S., almost 30% of all PET soft drink containers are recycled. In fact, PET is among the most recyclable materials and has the second highest scrap value after aluminum.

The technology of recycling is not new to the plastics industry. In fact, it has been practiced for many years in the form of internal recycle of waste, trim, etc. Frequently, market prices for plastics and the parts made from them could not be maintained if scrap were not recycled. Last year, almost 600,000 t of industrial plastic waste were recycled in the U.S. In Du Pont, worldwide, at least 450,000 t/year of plastics and plastic feedstocks are being recycled every year. In addition, more than 100,000 t/year of plastics are reclaimed from industrial operations in 12 recycling centers.

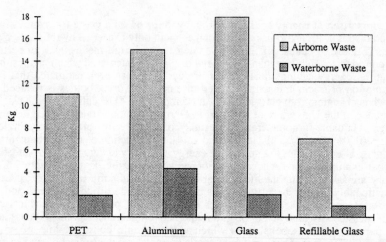

Figure 2. Environmental Impact of Soft Drink Containers for Delivery of 1000 L in 475 cc Containers (SOURCE: Adapted from Franklin Associates, Ltd.)

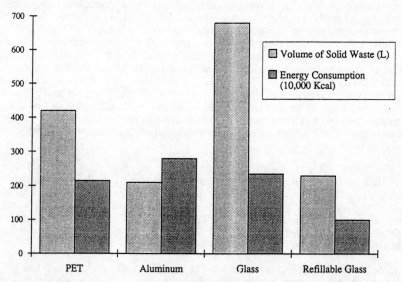

Figure 3. Environmental Impact of Soft Drink Containers for Delivery of 1000 L in 475 cc Containers (SOURCE: Adapted from Franklin Associates, Ltd.)

Industrial plastic waste usually consists of only one polymer type or grade and is relatively free of contamination. The majority is based on thermoplastic polymers which lend themselves readily to recycling. Because of their high stability, properties do not decline during recycling, and grinding often is the only operation needed before the material is ready to be melt processed into a finished product again. Often, the recycled material is melt blended with virgin material without any negative impact on properties or processing conditions.

All plastics can be recycled by some method. A hierarchy of recyclability (Fig. 6) shows, however, that the effort needed for recycling and the value of the reclaimed products is not the same in all cases. Condensation polymers such as nylon and polyesters offer the widest range of recycling approaches. For instance, in addition to re-use of the polymers by re-melting in modified or unmodified form, it is possible to recover pure monomers from highly contaminated polymer by chemical degradation. Several companies have announced processes for the methanolysis of waste PET which give back pure starting material for the preparation of high quality PET. This should facilitate recycling into the same end use even when food contact is involved. Resins with lower melt stability such as cast acrylics may be recycled only by low temperature cracking to the monomers. In the case of polyurethanes, digestion with water or alcohols results in lower molecular weight, melt-processible oligomers or monomers which can be used again as raw materials for plastics. Thermosets and vulcanized rubbers are considerably more difficult to recycle and may only find re-use as fillers. Thermoplastics may be preferred in the future because several recycling options exist which lead to high value products.

Recycling of post-consumer plastics has begun only recently and was initially limited to rigid packaging materials, especially PET bottles. The problems facing the recycler are many but are primarily in the areas of collection, sorting, and purification.

The plastic content of municipal solid waste in the U.S. is about 7.5 weight % which corresponds, according to recent studies, to about 18 volume %. Some 11 MM t of plastics go into municipal solid waste each year. Because of their short life cycle, packaging materials represent a high proportion of the total plastic component. Post-consumer plastic recycling in 1990 amounted to 310,000 t, a 45% increase over 1989 but still less than 3% of the total. About 82,000 t were PET soft drink containers. The existence of a collection infra-structure is essential in order to maintain a continuous stream of raw waste plastics for a recycling facility. Curbside collection programs have more than doubled since 1989 and at last count numbered over 1,000. Transportation costs can be a major impediment to economical recycling as is the case with foamed materials, i.e. foamed polystyrene or polyurethane. By contrast, plastic bottles and cans can readily be compacted to a density of at least 0.5 g/cc.

The variety of different materials, often used in one product, is another problem. For example, PET bottles may be clear or colored and can contain as many as 5 additional materials, i.e. a HDPE base cup, a paper or polypropylene label, an adhesive, and a metal or polypropylene cap. HDPE bottles range from clear milk bottles to highly pigmented liquid detergent bottles. Frequently, these materials arrive at the recycling plant commingled and pressed into bales and automated sorting facilities will be needed to separate these mixtures by polymer type and by color in an economic manner.

The value of clear, unpigmented resins is highest because it gives the user the widest range of application choices. For instance, clear PET can be spun into fiberfill or carpet fiber or with certain additives used as injection molding resin while pigmented PET may only be useful as inner layer in a multilayer product.

Packaging Material	Glass, Refillable	Paper Carton	PE Bag
Amount (g)	457	28	7
% of Total Weight	31	2.6	0.7

Figure 4. Packaging Material vs. Contents (1L Milk Containers) (SOURCE: Adapted from K. C. Domke, Bosch GmbH, Germany)

	Recycling Rate, %	Content in MSW, %
Aluminum	32	1
Paper	26	40
Glass	12	7
Steel	6	7
Plastics	1-2	7.5

Figure 5. Recycling Rates of Materials in Municipal Solid Waste (U. S.) (SOURCE: Adapted from Council for Solid Waste Solutions)

	Cond. Poly. (PET, nylon, TPE)	Polyolefins (HDPE, PP, TPO)	PVC ABS	Acrylics	Thermo-sets	Cured Rubbers
Direct re-melt	✓	✓				
Re-melt after modification	✓	✓	✓	(✓)		
Chemical monomer recovery	✓				(✓)	
Thermal monomer recovery		(✓)		✓		
Thermal cracking	✓	✓	(✓)	✓	✓	✓
Use as filler	✓	✓	✓	✓	✓	✓

Figure 6. Hierarchy of Recycling

The materials recovered from recycle operations can not automatically go back into the same end-uses. Concern over residual chemical contaminations, justified or not, will probably limit applications in the food container market. As mentioned earlier, this may be helped by "chemical," also called tertiary, recycling to the monomers, followed by re-polymerization. In the case of PET, purity of the resulting plastic is comparable to virgin material. Coca Cola, working with Hoechst-Celanese, has won FDA approval for re-use of this material in food contact applications. HDPE recovered from milk bottles often contains small amounts of fatty acids detectable by a slight odor and discoloration. Contamination with paper, metal, other polymers, and pigments, even if reduced to a very small level, can have considerable impact on the appearance of blow molded products. The presence of small amounts of plastics with different melting points and thermal stability can be a significant problem. Thus, PVC contamination of PET leads to discoloration while PET remains unmelted at PVC processing temperature. In both cases, part appearance is affected. For non-transparent containers, a multilayer approach may be feasible in which the recycled plastic is used for the inner layer only. Injection molding is less sensitive to traces of impurities but recycled plastics currently are mostly high melt viscosity blow molding resins. We have demonstrated with several injection molded products that up to 50% of recycled blow molding grade HDPE can be added to virgin plastic without affecting molding conditions or product properties. In a case like this, the recycled plastic becomes equivalent in value to the virgin plastic.

Injection molded bottle grade PET, because it lacks the orientation which blow molding provides, is too brittle for many applications. We have successfully used toughening technology to upgrade the value of recycled PET. For example, our modified recycled PET is an attractive material for pallets which have to withstand high loads at high temperatures. Several applications as containers and for automotive parts are presently under investigation. Substitution for lower cost plastics such as ABS is feasible with appropriate blending technology. Recycled HDPE has been used to demonstrate the fabrication of highway signs, milk bottle carriers, and waste containers of up to 3 m^3 in size.

Recycling of glass-reinforced plastics leads to a decrease in fiber length and some property reduction. Blends and alloys can be recycled just as easily as the individual components. We have shown that even multi-layered bottles or cans can be recycled into high value materials. It is sometimes necessary to adjust the ratio of the blend components to suit the product to the new use. In collaboration with several bottle and food producers, we have converted multi-layer polypropylene ketchup bottles into automobile bumper fascias. They met all specifications for bumper fascias from thermoplastic polyolefins and could be painted under currently used conditions.

More than 50% of the plastics in the municipal solid waste stream are not readily separated from each other. Polymers usually are incompatible with each other, and form separate phases when melted together, leading to a coarse-grained material with poor properties. Addition of compatibilizers can give some improvement but the ever changing composition of the post-consumer plastic waste stream makes compatibilization a difficult task. Nevertheless, commingled plastics have been used as wood replacement to make park benches, fence posts, etc. They have good weatherability but at equal volume, they are almost twice as heavy as wood. To cover recycling costs, these products may have to be sold for 2-3 times the price of wood.

In the durable goods industry, the average value of plastics tends to be higher than in packaging and, therefore, the economic incentive for recycling higher. However, the much longer use time and the often very harsh use conditions raise questions about loss of properties.

Several years ago, we examined this question for a number of nylon automotive parts and came up with very encouraging results (Fig. 7). It is this demonstrated stability that puts recycling of thermoplastic materials like nylon on a sound technical basis. As is the case with packaging materials, one of the biggest issues facing recycling of plastics from durable goods is collection, sorting and separation. Fortunately, collection of many durables is much more centralized and a significant amount of disassembling and sorting of nonplastic parts is already being practiced. It should be possible to extend this to plastics, especially if the number of different polymer types used by manufacturers is reduced somewhat and disassembly is already taken into consideration at the design stage. Labeling of plastic parts by a standardized method would also facilitate recycling. Efforts in all of these areas are already underway. On the other hand, once the multitude of plastic components of an automobile has been reduced to shredder waste, recycling is exceedingly difficult and costly and chances for returning a valuable product to the market are small.

An example for the future direction of automotive plastics recycling is legislation proposed in Germany. It would require car manufacturers, dealers, or their designates to take back used automobiles from the last user, starting 12/31/93. The goal is to achieve environmentally acceptable disposal of the entire car. Initially, 25% of the plastic components of new cars must be made from recycled materials. The German automobile industry has responded by proposing to set up a network of licensed disassemblers who will return all car components to the appropriate raw materials producers. Several car manufacturers, e.g. VW and BMW have set up model disassembly facilities to study the dismantling of their cars and to train disassemblers. During the last 8 months, the VW disassembly facility has removed over 50,000 bumper fascias from old cars and recycled them into fascias for new cars. If recycling into the same part is to be extended to many or all plastic components of an automobile, a significant reduction in the number of plastic types has to occur. Otherwise, individual recycle streams will be too small for economical operation. Even then, the variety of different grades of the same plastic type which will have to be processed together will pose a challenge to the capabilities of the recycler.

Where recycling is not feasible for technical or economic reasons, incineration is the best alternative. It recovers most of the fuel value originally borrowed from crude oil (Fig. 8). In fact, the high energy content of plastics facilitates incineration of municipal solid waste with high content of water and non-combustible materials. Today, incinerators are operated all over the world and it has been amply demonstrated that proper flame temperature permits safe operation and removes the risk of dioxin formation. Tests have shown that burning polyvinyl chloride in properly operated incinerators does not lead to dioxins in the off-gases. Together, all 47 incinerators currently operating in Germany can by law not produce more than 400 g dioxins per year. A recent change of the law will reduce this to 4 g per year! Some recent data suggest that dioxin formation in home heating systems, especially when wood fired, exceeds by several orders of magnitude those in modern waste incinerators. Concerns over the disposal of residual ash from plastic-burning incinerators will be reduced once heavy metal-containing pigments and stabilizers are deleted from plastics.

Much has been said recently about degradable plastics. Some have suggested that this would be the ultimate solution to the plastics disposal problem while others have rejected this approach as incompatible with recycling. There is no doubt that after years of raising the stability of plastics to meet ever increasing end-use demands, the notion of manufacturing degradable plastics seems difficult to accept. Further, it is

Parts tested:
        Fluid reservoirs
            Brake fluid
            Power steering fluid
        Stone shield
            Disc brake
            Steering column
        Engine flex fan
        Transmission downshift cable
        Emissions canister
        License plate pocket
        Rear body panel

Service life:
        50,000-160,000 km (2-10 years)

Results:
        • Good retention of resin quality
        • No adverse effects of environmental conditions

Figure 7. Field Perfomance of Automotive Nylon Parts (from serial production) (SOURCE: Adapted from H. H. Goodman, SAE Technical Series 830076)

Material	Btu/Pound
Plastics	
• Polyethylene	19,900
• Polypropylene	19,850
• Polystyrene	17,800
Rubber	10,900
Newspaper	8,000
Leather	7,200
Corrugated Boxes (paper)	7,000
Textiles	6,900
Wood	6,700
Average for Municipal Solid Waste	4,500
Yard Wastes	3,000
Food Wastes	2,600

Figure 8. Energy Values of Materials (SOURCE: Adapted from Council for Solid Waste Solutions)

now well known that little degradation of any substance occurs in landfills because of lack of light and air, and of sufficient moisture and nutrients to sustain microbial action. Thus, degradable plastics are not the answer to the landfill problem until an infrastructure of managed degradation plants, e.g. composting facilities, is in place. The number of communities with composting facilities is increasing, however, and consumer awareness of the need to segregate components of household waste is growing. It is difficult to see how degradable plastics could be introduced into durable goods unless degradation can be triggered with a signal which is not normally present in the environment. Alternatively, an effective but readily removable protection against premature degradation must be found. Degradable plastics will probably have a significant role in certain packaging, agricultural, and marine applications where intended product life is fairly short.

RECEIVED March 9, 1992

Chapter 3

Recycled Plastics

Product Applications and Potential

Robert A. Bennett

College of Engineering, University of Toledo, Toledo, OH 43606–3390

Plastics recycling requires information in four areas. These areas are 1) collection of plastics, 2) separation, 3) reprocessing technology, and 4) markets for recycled products. The fourth area is where this research was performed. Available markets will assure that post consumer polymer scrap has an economic value. Additionally, an economically viable market for recycled plastic products has the social benefit of reducing the flow of plastics into the solid waste stream. Increased plastics recycling will assist the U.S. Environmental Protection Agency in achieving the 1992 goal of either reducing or recycling solid waste by 25%.

Our research results found that approximately 190 million pounds of PET beverage containers were collected for recycling in 1989. Approximately 145 million pounds of HDPE, 20 million pounds of polystyrene (PS), 60 million pounds of polypropylene (PP), 5 million pounds of polyvinyl chloride (PVC) and 30 million pounds of mixed commingled plastics were reported to have been recycled in 1989. Recycling information was determined from a national survey of active plastics recyclers. Projections of market potential for recycled resins are presented. These projection estimates indicate that potential demand for recycled resins is much greater than supply.

Manufacturers of plastic pallets were contacted to determine interest in utilizing recycled plastics in their products. Questionnaires were sent to 82 plastic pallet companies across the country. Survey results showed that approximately 350 million pounds of virgin HDPE were reported to be used in pallet manufacturing. Reprocessed HDPE from plant scrap and post consumer scrap was reported to be a source for another 30 million pounds of HDPE.

The objective of this research was directed at gaining an improved understanding of new product opportunities, markets and the economics associated with the emerging plastics recycling business. Goals include investigating potential new products and markets for recycled plastics and developing an electronic database on

0097–6156/92/0513–0026$06.00/0

the plastics recycling industry. Research was performed in conjunction with other plastics recycling research projects funded through The Plastics Recycling Foundation/Center for Plastics Recycling Research.

Plastics recycling requires information in four areas. These areas are 1) collection of plastics, 2) separation, 3) reprocessing technology, and 4) markets for recycled products. The fourth area is where this research was performed. Available markets will assure that post consumer polymer scrap has an economic value. Additionally, an economically viable market for recycled plastic products has the social benefit of reducing the flow of plastics into the solid waste stream. Increased plastics recycling will assist the U.S. Environmental Protection Agency in achieving the 1992 goal of either reducing or recycling solid waste by 25% *(1)*. Table 1 defines the goals for the 1992 reduction and recycling of solid waste. The current situation is that landfilling is utilized for approximately 80 percent of solid waste disposal.

Table 1   Disposal   Goals for MSW

Method	Currently	1992 Goal
Landfill	80%	55%
Incinerate	9%	20%
Recycling	11%	25%

Source: U.S. EPA

Supply

Approximately 66 billion pounds of plastics were sold in the U.S. during 1989 (Table 2). Sales data were obtained from Modern Plastics *(2)* and the Fiber Organon *(3)*. Approximately 1.6 billion pounds of PP may have been double counted since some PP fiber is included in PP non textile sales.

Polyethylene (high and low density) remained the dominant resin with a total of over 18 billion pounds sold which represents about 32% of the total nontextile plastics resin sales. Polyester resin for nontextile products accounted for 2 billion pounds or 3.6% of total plastics sales. Polyester used in textiles accounted for another 3.5 billion pounds bringing both uses of PET to total over 9% of total plastic sales. Polyethylene and polyester are the dominant plastics currently used in recycling post consumer plastics. Approximately 85% of plastic bottles are made from these plastics. Post consumer scrap generated from these billions of pounds of thermoplastics offer recycling opportunities.

Reuse of thermoplastics will reduce raw material costs to manufacturers and reduce the burden caused by plastics on the solid waste stream.

Analysis of the national survey results obtained as part of this research shows that established PET and HDPE recyclers are experiencing a continued supply shortage of post consumer plastic scrap. Many recyclers are advertising in trade journals for additional scrap. Major corporations have entered into the plastics recycling business. E.I. du Pont de Nemours & Co. and Waste Management, Inc. have formed a joint venture that will sort and recycle plastics from municipal solid waste *(4)*. It is estimated by du Pont that a new PET resin plant would cost between $1.25 and $1.50 per annual pound to build while a recycled resin plant is about 15 cents per pound to build. This type of economic incentive makes recycling plastics an attractive investment.

Table 2 - U.S. Plastics Sales 1989

Material	1989	Million lbs.
ABS		1243
Acrylic		739
Alkyd		325
Cellulosics		91
Epoxy		492
Nylon		595
Phenolic		3162
Polyacetal		141
Polycarbonate		622
Polyester, thermoplastic (PBT, PCT, PET)		2101
Polyester, unsaturated		1325
Polyethylene, high density		8115
Polyethylene, low density		10636
Polyphenylene-based alloys		176
Polypropylene and copolymers		7246
Polystyrene		5184
Other styrenics		1170
Polyurethane		3245
Polyvinyl chloride and copolymers		8307
Other vinyls		960
Styrene acrylonitrile (SAN)		137
Thermoplastic elastomers		542
Urea and melamine		1367
Others		307
Sub Total		**58,228**

Textile Fibers		
Yarn and Monofilaments		
Nylon - Industrial		364.6
- Carpet		967.7
- Textile		400.4
Polyester - Industrial		365.0
- Textile		804.0
Olefin		1,251.0
Total Filament Yarn		**4,152.7**

Staple & Tow & Fiberfill		
Nylon		978.5
Acrylic - Modarcylic		537.2
Polyester		2,353.5
Olefin		375.4
Total Staple & Tow & Fiberfill		**4,244.6**
Total Textile Fibers*		**8,397.3**

Total U.S. Sales		**66,625.3**

*Textile totals include domestic and export shipments.
Source of data:
    Modern Plastics, Jan.,1990; Fiber Organon, Jan., 1990

Other plastics recycling ventures include Clean Tech, Inc., which has been created by Proctor & Gamble and Plastipak Packaging. Clean Tech will recycle PET & HDPE bottles. Located in Dundee, Michigan, this new company will provide a long term supply of recycled resins to be used by Plastipak in producing packaging. Proctor & Gamble currently markets some of its household products in containers made from recycled materials *(5)*. Wellman, Inc., the largest plastics recycler, also has announced a joint venture with Browning Ferris, Inc. (BFI), a national waste hauler. Cooperative ventures will open channels to capture post consumer plastics before these materials become part of the solid waste stream. Currently the major source of plastics collection is mandatory deposits in certain states. By utilizing the solid waste stream many more types of containers and varieties of plastics can be reclaimed for recycling. Numerous suppliers of commercial (turnkey) recycling facilities have recognized the potential for reclamation of plastic waste.

Plastics recycling has a large growth potential and is drawing primary and auxiliary machine suppliers to focus their R&D and marketing efforts toward this area. The development of off-the-shelf recycling systems is the goal of most of these companies. The off-the-shelf recycling systems will reclaim a variety of plastic scrap. These systems sort, shred, clean, and pelletize the plastic scrap. Many companies are modifying their machinery in growing numbers to handle post-consumer scrap. Equipment suppliers are seeking to license recycling technology to broaden applications for their systems *(6)*. A number of U.S. based companies supply total recycling systems or "turnkey plants". These firms either have backgrounds in material size reduction such as baling or grinding, conveying, and/or pelletizing equipment, or have expertise in recycling metals and glass *(7)*. A listing of some suppliers of turnkey plants are:

Automated Recycling Corp., Sarasota FL
Conair Reclaim Technologies, Pennington, NJ
John Brown Plastics Machinery, Providence, RI
National Recovery Technologies, Nashville, TN
Pure Tech International, Pine Brook, NJ
Recycled Polymers, Madison Heights, MI

Presently there are fifteen large plastics recycling plants operating in the United States plus a number of smaller facilities. The recycling capacity of the fifteen plants is approximately 534 million pounds of plastics *(7)*. Table 3 shows the capacity of these larger plastics recycling plants.

Quantum Chemical Corp., a resin producer, also announced that it has entered into the plastics recycling business. It plans to build a 40 million pounds per year rigid container recycling facility next to an existing compounding plant in Heath, Ohio. It will process post-consumer HDPE, PET, and PP containers and LDPE film *(8)*. Sunoco Products and Mindis Recycling, Inc., are planning a plastics recycling facility capable of processing over 200 million pounds annually. This would be the largest such plant in North America. The plant is to be located in Atlanta, Georgia and will have five processing lines to handle HDPE film, HDPE bottles, PET and polystyrene. The companies plan to begin operation in December. The plant will be a 108,000 square feet building and will utilize a variety of components rather than a single company's line of recycling equipment. An estimated $23 million is being budgeted for this project and it will employ approximately 60 workers *(9)*. This additional capacity of Quantum Chemical and the Sunoco-Mindis project will bring total recycling plant capacity to 774 million pounds.

Table 3 Capacity of U.S. Plastics Recycling Plants

Company	Capacity*	Resin
Clean Tech	12	HDPE PET
Day Products	40	PET
Eaglebrook	35	HDPE
Graham Recycling	20	HDPE
Johnson Controls	20	PET
M.A. Industries	NA	HDPE PET Mixed HDPE/PET all Mixed Rigid Containers
Midwest Plastics	10	HDPE
Orion Pacific	7	Mixed HDPE/PET
Partek	5	HDPE
Pelo Plastique	15	PET
Plastic Recycling Alliance (incl 2 plant sites)	80	Mixed HDPE/PET all Mixed Rigid Containers
Star Plastics	32	HDPE PET Mixed HDPE/PET
St. Jude	25	PET
Union Carbide	50	Mixed HDPE/PET
United Resource Recovery	8	HDPE
Wellman	175	PET Mixed HDPE/PET
Total	**534**	

* Capacity in Million Pounds

Source: Resource Recycling Plastics Recycling Update

PET Container Recycling

Growth of PET container recycling is continuing. Cost advantages and reduced investment realized by using reclaimed plastics rather than using or manufacturing virgin plastics will be a driving force in increasing the volume of recycled plastics. Current research estimates that approximately 190 million pounds of PET soft drink containers were recycled in 1989. Figure 1 shows the rapid increase in recycling plastic beverage containers.

Major markets for recycled PET continue to be in carpeting, fiberfill, unsaturated polyester, polyols for rigid urethane foam, strapping, engineering plastics and extruded products, see Figure 2.

New applications such as thermoformed products and textiles/geotextiles offer additional opportunities. Companies such as Hoechst Celanese, Goodyear's Polyester Division *(10)* and Eastman Chemical Division *(11)* have been working on depolymerizing PET then repolymerizing to obtain a purified resin. Goodyear has introduced a new recycled resin called Repete while Eastman has developed a methanolysis process for converting scrap PET to PET by repolymerization. The

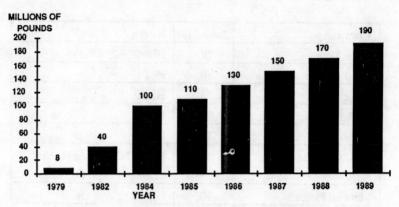

Figure 1 PET Beverage Container Recycling

Figure 2 Potential Markets for Recycled PET

major soft drink companies, Coca-Cola Co. and Pepsico, Inc., intend to utilize recycled PET in soft drink containers *(12)*.

In 1989, there were 680 million pounds of PET used in manufacturing soft drink containers. Estimates show that there exists a potential market for approximately 530 million pounds of this material in non food applications *(13)*. This market projection of 530 million pounds is based upon estimates of recycled plastic utilization for various product applications, see Table 4. Actual sales data were obtained from sources such as Modern Plastics, *(2)* and the Fiber Organon *(3)*. Potential market size was developed by assuming no food contact application and today's available technology. It is believed that the potential market size of 530 million pounds is relatively conservative since food applications are now beginning to be realized *(12)* and process modifications are available to further increase usage of recycled PET. Since an estimated 190 million pounds are only being recycled, this market is far from being saturated.

Table 4 PET Potential Market

POLYETHYLENE TEREPHTHLATE - PET- Recycling Market Projections

Major Markets [million lbs.]	Actual 1989 Use	Potential Market Size Percent Penetration	Resulting Volume
Polyester Thermoplastic (PET)			
Blow molding			
Soft-drink Bottles	680	0%	0
Custom Bottles	260	10%	26
Extrusion			
Film	520	10%	52
Magnetic Recording Film	81	0%	0
Ovenable Trays	39	0%	0
Coating for ovenable board	10	0%	0
Sheeting (for blisters,etc)	70	10%	7
Strapping	32	40%	13
Exports	213	10%	21
Total PET	1,905		119
Polyester, Unsaturated			
Reinforced Polyester			
Molded	783	2%	16
Sheet	165	2%	3
Surface Coating	20	0%	0
Export	30	0%	0
Other	327	0%	0
Total Unsaturated Polyester	1,325		19
Reinforced Polyester; Unsaturated			
Aircraft/aerospace	35	0%	0
Appliance/business	93	2%	2
Construction	426	2%	9
Consumer	138	0%	0
Corrosion	350	0%	0
Electrical	53	0%	0
Marine	353	2%	7
Transportation	221	2%	4
Other	52	0%	0
Total Reinforced Unsat. Poly.	1,721		22
Polyurethane - Rigid Foams			
Building Insulation	450	2%	9
Refrigeration	160	2%	3
Industrial Insulation	90	2%	2
Packaging	68	2%	1
Transporation	48	2%	1
Other	47	0%	0
Total Polyurethane	863		16
Textile			
Filiment Yarn	1169	10%	117
Staple and Tow	2354	10%	235
Total Textile	3,523		352
GRAND TOTAL	9,337		529
1995 PROJECTED GRAND TOTAL at 3% growth rate	11,149		631

HDPE Recycling

HDPE sales of virgin resin in 1989 were in excess of 8.1 billion pounds, see Table 2. About 145 million pounds of HDPE have been identified as being recycled in 1989. This represents a significant increase of over 55% from the approximately 93 million pounds in 1988. Many communities are beginning to collect milk jugs and water jugs and detergent containers for HDPE recycling. The growth in HDPE recycling is presented in Figure 3.

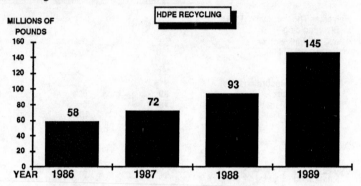

Figure 3 HDPE Recycling

AGRICULTURE	GARDENING
Drain Pipes	Flower Pots
Pig and Calf Pens	Garden Furniture
	Golf Bag Liners
MARINE ENGINEERING	Lumber
Boat Piers (lumber)	
	INDUSTRIAL
CIVIL ENGINEERING	Drums/Pails
Building Products	Kitchen Drain Boards
Curb Stops	Drums/Pails
Pipe	Kitchen Drain Boards
Signs	Matting
Traffic-Barrier Cones	Milk Bottle Carriers
	Pallets
RECREATIONAL	Soft Drink Base Cups
Toys	Trash Cans
	Household Chemicals

Figure 4 Potential Markets for Recycled HDPE

Products for recycled HDPE are soft drink basecups, plastics pipes, plastic lumber, and various containers including household chemical containers, see Figure 4.

A potential market of an estimated 442 million pounds could be developed to utilize recycled HDPE, see Table 5. This market potential estimate was calculated in a similar manner as the PET estimate. The 145 million pounds of HDPE being recycled represents less than a third of this market potential identified for recycled HDPE for non food applications.

Table 5 Recycled HDPE Potential Market Projections

Major Markets [million lbs.]	Actual 1989 Use	Potential Market Size Percent Penetration	Resulting Volume
Blow Molding			
Bottles			
Milk	720	0%	0
Other Food	320	0%	0
Household Chemicals	956	10%	96
Pharmaceuticals	208	0%	0
Drums (>15 gal.)	154	5%	8
Fuel Tanks	96	0%	0
Tight-Head Pails	90	10%	9
Toys	70	5%	4
Housewares	51	0%	0
Other Blow Molding	270	0%	0
Total Blow Molding	**2,935**		**116**
Extrusion			
Coating	51	0%	0
Film (< 12 mil.)			
Merchandise Bags	182	0%	0
Tee-shirt Sacks	215	0%	0
Trash Bags			
Institutional	124	0%	0
Consumer	15	0%	0
Food Packaging	94	0%	0
Deli Paper	16	0%	0
Multiwall Sack Liners	50	0%	0
Other	70	0%	0
Pipe			
Corrugated	103	25%	26
Water	59	0%	0
Oil & Gas production	70	0%	0
Industrial/Mining	61	0%	0
Gas	112	0%	0
Irrigation	40	50%	20
Other	46	0%	0
Sheet (> 12 mil)	305	10%	31
Wire & Cable	146	0%	0
Other Extrusion	36	10%	4
Total Extrusion	**1,795**		**80**
Injection Molding			
Industrial Containers			
Dairy Crates	56	10%	6
Other Crates, Cases, Pallets	120	10%	12
Pails	380	10%	38
Consumer Packaging			
Milk-bottle Caps	25	0%	0
Other Caps	56	0%	0
Dairy Tubs	136	0%	0
Ice-cream Containers	85	0%	0
Beverage-bottle Bases	122	50%	61
Other Food Containers	46	0%	0
Paint Cans	31	10%	3
Housewares	190	0%	0
Toys	78	5%	4
Other Injection	218	10%	22
Total Injection Molding	**1,543**		**145**
Rotomolding	122	10%	12
Export	830	0%	0
Other	890	10%	89
GRAND TOTAL	**8,115**		**442**
1995 PROJECTED GRAND TOTAL at 3% growth rate	**9,690**		**528**

Polypropylene (PP) Recycling

PP sales of virgin resin in 1989 were in excess of 7.2 billion pounds, see Table 2.
About 60 million pounds of PP have been identified as being recycled in 1989.
Topping the list of products identified as being manufactured from recycled resin is
the automotive battery case. This product has already proven to be a success in the
recycling area. Products for recycled PP are listed in Figure 5.

Auto parts (batteries)	Bag dispensers
Bird feeders	Flower pots
Furniture	Golf equipment
Pails	Pallets
Water meter boxes	Carpet
Slip Sheets	

Figure 5   Potential Markets for recycled PP

Growth of recycled PP is expected in the textile fiber industry, pails, pots,
and automotive applications. The 62 million pounds of PP recycled represent less
than 10% of the estimated potential market for recycled PP.

Polyvinyl Chloride (PVC) Recycling

Total sales of PVC in 1989 amounted to more than of 8.3 billion pounds (refer to
Table 2). Currently, approximately 5 million pounds or 0.06% has been identified
as being recycled by consumers. Products not related to the food or health care
industries make excellent candidates for manufacturing from recycled (reprocessed)
resin. A potential market list is given in Figure 6.

Downspouts	Bottles
Fencing and corrals	Flower pot covers
Handrails	Flower/bud vases
House siding	Garden hose core
Landscape timbers	Office accessories
Sewer/drain pipe	Stadium bleachercovers
Telephone cables	Truck bed liners
Urinal drain covers	Floor/machinery mats
Window frames	Refuse containers

Figure 6 Potential Markets for Recycled PVC

Market analysis shows that a potential market of an estimated 494 million
pounds could be developed to utilize recycled PVC. Major potential markets for
recycled PVC are flooring, hoses, new bottles and various types of drain and sewer
pipes. Considerable amounts of PVC are recycled as in-plant scrap. However,
only about 5 million pounds of post consumer PVC are being recycled as post
consumer scrap and this represents only about 1% of the potential market for PVC
recycling.

Polystyrene (PS) Recycling

In 1989, the total sales of PS accounted for more than 5.2 billion pounds (refer to
Table 2). Of this amount, approximately 20 million pounds or less than 0.4% has

been identified as being recycled by consumers. Some potential products to be manufactured from recycled PS resin include insulation board, appliance housings, and various trays. Research shows that a potential market of an estimated 477 million pounds could be developed to utilize recycled PS. Since only 20 million pounds of PS were reported as being recycled, this represents less than 5% of the potential market.

Commingled Plastics

Plastics recycled from the solid waste stream often result in a mixture of many types of plastics. Separation of this mixture initially is currently limited to gleaning out the easily recognized containers such as soft drink, milk and detergent bottles. The remaining mixed commingled plastics can be manufactured into noncritical products. A listing of products for recycled mixed commingled plastics include outdoor fencing, pens, benches and picnic tables. Applications of mixed commingled plastics for plastic pallets offers another product opportunity.

Recycled Plastics for Pallets

Questionnaires were sent to 82 plastic pallet companies across the country. Almost 350 million pounds of virgin HDPE were reported to be used in pallet manufacturing. Calculations show that for every 1% penetration of recycled plastics into the wooden pallet industry, approximately 370 million pounds of plastics could be utilized. The National Wooden Pallet and Container Association (NWPCA) in Washington, D.C., estimates new pallet production at over 465 million. This large market offers considerable opportunities for utilizing recycled plastics. Superwood International, Ltd. of Wicklow, Ireland, identified the manufacturing of pallets from recycled material as a major niche market for its production technology *(14)*. Their process is best suited for recycling industrial scrap. Feed material could include HDPE, LDPE, PP and even wood chips as filler.

Product Modeling

A research technique utilized to assist in designing new product applications for recycled plastics is the method of finite elements. The finite element method (FEM) is a technique that divides a sample into sufficiently small regions so that the solution in each small region (element) can be represented by a simple function of displacement and stress. This technique has applications in a broad range of physical problems. The subregions or elements are joined together mathematically by enforcing conditions that make each element boundary compatible with each of its neighbors while satisfying the region boundary requirements. This method was used to study the distribution of stresses on the pallet and to improve the existing design. An exaggerated view of a loaded pallet as generated by the computer model is shown in Figure 7. This model suggests that the thickness of the plastic pallet should be approximately 30% greater than a wooden pallet to perform as well. This modeling technique reveals that plastic pallets should be redesigned to take advantage of the properties of plastics. The plastic boards in pallets could for example be extruded in an I-beam cross section to increase strength while minimizing weight.

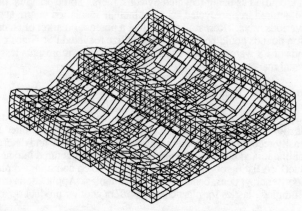

Figure 7   Modeled plastics pallet (exaggerated view)

Plastics Recycling Future

Plastics recycling is facing increasing pressure due to dwindling supply of landfill space and legislation to reduce the solid waste stream. Increasing the amount of recycled plastics in the future will require improved high speed sorting and novel collection techniques. Further research in product and market development will be required to maximize the value and usage of recycled plastics.

Acknowledgments

Research was performed in conjunction with other plastics recycling research projects funded through The Plastics Recycling Foundation/Center for Plastics Recycling Research. Additional funding was provided by the Ohio Department of Natural Resources, Division of Litter Prevention & Recycling, Edison Polymer Innovation Corporation and The University of Toledo. The assistance of the following graduate students is greatly appreciated: Mr. Chu-Cheng Khor, Mr. Bitan Banerjee, Ms. Debbra Lefcheck and Mr. B. T. Srinath.

References

1. U.S. EPA, The Solid Waste Dilemma: An Agenda for Action (1989).
2. "Resins 1990", *Modern Plastics* , January, 1990, pp. 99-108.
3. *Fiber Organon* , January, 1990, pp. 18.
4. B. Richards, A. K. Naj, "Du Pont and Waste Management Plan To Build Largest U.S. Recycling Plant", *Wall Street Journal*, April 26, 1989, pp. A-4.
5. "Plastics Recycling Ventures Started", *Chemical & Engineering News* , December 4, 1989, pp. 14.
6. P. A. Toensmeier, "Solid-waste crisis creates market for dedicated recycling lines", *Modern Plastics* , 1989, pp. 62.
7. R. V. Wilder, "How sound an investment are turnkey recycling plants?", *Modern Plastics* , July, 1990, pp. 45-47.
8. "Recycling Processor Update", *Resource Recycling's Plastics Recycling Update* , February, 1991, pp. 3.
9. D. Loepp, "Sonoco, Mindis Plan Huge Recycling Plant", *Plastics News* , July 1, 1991, pp. 1.

10. S. Combs, "Coke, Pepsi to Recycle PET Bottles to Bottles". *Waste Age's Recycling Times*, December 18, 1990, pp. 1.
11. K. Meade, "Eastman Tries Recycling Process For PET in Food Applications", *Waste Age's Recycling Times*, December 18, 1990, pp. 3.
12. J. Powell, "It's the real thing: the recycled plastic soft drink bottle", *Resource Recycling*, January, 1991, pp. 37-39.
13. R. A. Bennett, Technical Report #41, Center for Plastics Recycling Research, "New Product Applications and Evaluation & Continued Expansion of a National Database on the Plastics Recycling Industry", (1990).
14. "Recycling System to be Licensed for Making Pallets", *Plastics Technology*, March, 1989, pp. 41.

RECEIVED April 28, 1992

Chapter 4

Miscibility in Polymer Recycling

Richard S. Stein

University of Massachusetts, Amherst, MA 01003

Thermodynamic principles governing miscibility are discussed, and the reason for most mixtures encountered in recycle feedstock being immiscible are explained. The relationship between miscibility, interface diffuseness and mechanical strength are discussed as are means for their modification. Factors governing the crystallization of both miscible and immiscible polymer mixtures are analyzed. It is demonstrated that the properties of polymers prepared from mixed feedstocks can be affected by both composition and processing conditions.

Polymers which are recycled may be classified into categories of those which are:
1. Commingled
2. Separated

Commingled polymers are those where different species are mixed together, whereas separated ones have these separated. Separated plastics are most valuable for recycling since their properties are not degraded as a result of mixing them with other plastics. The problem is that sources of separated plastics are limited. These arise from:
1. Plastic waste arising from manufacturing processes.
2. Plastic articles separated by the consumer
3. Plastic articles separated after collection from the consumer

An appreciable amount of "in house" recycling takes place in Category (1), where scrap resulting from the fabrication of plastic articles is collected and reused by the manufacturer. This is a process which has been taking place for many years and it will undoubtedly continue to increase. It is cost effective for the manufacturer and processor.

For Category (2), it is necessary that the consumer be able to recognize plastics of different types and conveniently return these to some collector. It has been successful, for example, for polyethylene terephthalate (PET) soft drink bottles, where return can be encouraged (in some localities) by imposing a deposit, and redemption machines have been installed in supermarkets which read bar codes on bottles, refund the deposit, and then grind them up for compactness in shipping. Another readily identifiable polymer is high density polyethylene (HDPE) used for

0097–6156/92/0513–0039$06.00/0
© 1992 American Chemical Society

milk and water jugs. It has been reported that 25% of the plastic jugs which Exxon uses for motor oil come from this source *(1)*.

Styrofoam is also readily recognizable and is used for coffee cups, hamburger "shells" and cafeteria trays. These are often used in centralized locations such as school cafeterias where collection is feasible.

Separated plastics can also come from larger scale users. For example, polyethylene from agricultural mulch and polypropylene from bailing straps. A problem that occurs with even separated plastics is that they can become mixed with small amounts of foreign plastics. For example, PET soda bottles often have polyethylene bases and polypropylene caps. It would be desirable to encourage manufacturing practices which avoid such mixing, as the necessary separation adds to recycling costs. Styrofoam products are often contaminated with paper and food wastes, but these can easily be removed by washing with water and detergent and floatation.

There are not too many examples in current use of where plastics are separated after collection from the consumer. Manual separation is labor intensive and expensive, and automatic means for this are not readily available. In principle, separation might be carried out by a device that could detect some property of the plastic, and separate it accordingly. The property could be a bar code, a spectrographic indicator, or an NMR signal. While there is need for research to develop such devices. it is uncertain whether such automatic separation can be accomplished economically.

Other than the above, a large fraction of the plastic waste stream consists of mixed plastics which cannot be easily or economically separated. One must then consider how to use these in forms where there properties will not seriously suffer.

Properties of Commingled Plastics

As will be discussed in the next section, most plastic mixtures occurring in commingled plastics will be "immiscible". That is, they will not form a single phase but will separate into two or more phases. These will be separated by boundaries which may be sharp or diffuse. With a sharp boundary between two different polymeric regions, there is often little molecular interpenetration, so that there is a region of mechanical weakness. Thus, failure is likely at these low adhesion boundaries, so the physical properties of such an immiscible commingled mixture will generally be poorer than those of the individual components.

To improve the properties of phase separated systems, it is therefore desirable to increase the strength of the interface. this may be accomplished by

1. Modifying one or both components by techniques such as grafting or copolymerization so as to render them more miscible.
2. Carrying out a chemical reaction so as to bind components together at the interface. The binding may be through chemical reaction, such as grafting, or else through secondary interactions such as hydrogen bonding or charge transfer.
3. Adding an interfacial agent which binds the two phases together. An example is a diblock coplymer, one block of which is miscible with one of the phases and the other block with the other phase. Such materials act like emulsifying agents to stabilize immiscible suspensions.

Principles of Miscibility

As mentioned, chemically different polymers are usually immiscible. It may be understood why immiscibility is more common for polymer pairs than for low molecular weight species in terms of thermodynamic considerations.

For a process to occur spontaneously at constant temperature, T, and pressure, P, it is necessary that the Gibbs free energy, G, decrease. For two polymers to dissolve in each other, this free energy change is the free energy of mixing, ΔG_{mix}, which may be resolved into enthalpy, ΔH_{mix}, and entropy, ΔS_{mix}, components according to

$$\Delta G_{mix} = \Delta H_{mix} - T \Delta S_{mix}$$

The entropy of mixing of an ideal solution is given by

$$\Delta S_{mix} = - R \ [\ n_1 \ ln \ x_1 + n_2 \ ln \ x_2 \ ]$$

where n_1 and n_2 are the numbers of moles of the two components and x_1 and x_2 are their mole fractions. Since the x_i terms are 1 or less, their logarithms will be zero or negative, so ΔS_{mix} is generally positive. This is reasonable since entropy is associated with disorder, and a solution is more disordered than the separated components. The increase in ΔS_{mix} contributes to a decrease in ΔG_{mix} and is a principal driving force for materials being soluble.

The above equation for ideal solutions presumes that molecules of both components are of equal size. This is usually not true for polymer mixtures, so refinements by Flory *(2)*, and others lead to the modification,

$$\Delta S_{mix} = - R \ [\ n_1 \ ln \ \phi_1 + n_2 \ ln \ \phi_2]$$

Here, the ϕ's are the volume fractions which become equal to the mole fractions when the molecules are of equal size. This modification also leads to zero or positive ΔS_{mix}.

The number of moles of component i is given by

$$n_i = w_i \ / \ M_i$$

where w_i is its weight and M_i is its molecular weight. For polymer mixtures, the M's are large so the n's are small. Thus, ΔS_{mix} will be less for polymer mixtures than for low molecular weight mixtures. The physical significance of this is that if monomer units are linked together to form a polymer chain, their are fewer ways for them to mix than if they are not. A consequence is that ΔS_{mix} is a smaller driving force for miscibility of polymers than for low molecular weight species so that miscible polymer pairs are less frequent. Consequently, polymer miscibility is more controlled by ΔH_{mix} than is the case for low molecular weight mixtures.

The Enthalpy of Mixing of Polymers

Theories of ΔH_{mix} based on nearest neighbor pair interactions have been proposed by Scatchard and Hildebrand *(3)*. These lead to equations of the sort

$$\Delta H_{mix} = RT \ \chi \ n_1 \ \phi_2$$

where χ is the "Flory interaction parameter" *(2)* which describes the molecular interactions. This may be related to the pair interaction potentials, ε_{11}, ε_{22}, and ε_{12} between molecular segments by

$$\chi = (z \, AV_1/RT)[\, (\varepsilon_{11} + \varepsilon_{22})/2 - \varepsilon_{12} \,]$$

where z is the coordination number of the lattice, A is Avogadeo's number and V_1 is the molar volume of component 1. In an <u>athermal</u> mixture, the 12 interaction is the average of the 11 and the 22, so

$$\varepsilon_{12} = (\, \varepsilon_{11} + e_{22}) / 2$$

so χ and ΔH_{mix} are both zero. In this case, ΔG_{mix} is completely determined by the ΔS_{mix} contribution and miscibility will ensue.

For van der Waals (dispersion) forces between segments, ε_{12} is the geometric mean of ε_{11} and ε_{22} (4),

$$\varepsilon_{12} = (\, \varepsilon_{11}\varepsilon_{22})^{1/2}$$

so

$$\chi = (z \, AV_1/RT) \, [\, (\, (\varepsilon_{11} + e_{22}) / 2 - \varepsilon_{11}\varepsilon_{22})^{1/2} \,]$$
$$= (z \, AV_1/RT) \, [(\varepsilon_{11})^{1/2} - (\varepsilon_{22})^{1/2}] \, ^2$$

This will always be a positive quantity so van der Waals interactions always leads to a positive H_{mix} contribution to ΔG_{mix}, which promotes miscibility. The $(\varepsilon_{tii})^{1/2}$ terms are related to the solubility parameters, δ_t, so one may write

$$\chi = (V_1/RT) \, [\, \delta_1 - \delta_2 \,]^2$$

The greater the difference between the δ's, the more positive the ΔH_{mix} and the less the miscibility. Thus, substances tend to be miscible if their solubility parameters are close together. This is a quantitative expression of the adage that "like dissolves like". It is noted that for constant ε's, this predicts that χ should vary inversely with T.

A negative ΔG_{mix} (and miscibility) can be achieved if ΔH_{mix} is small enough to be less than the small ΔS_{mix} characteristic of the molecular weights of the components. This will occur if the δ's are sufficiently close together. For a particular molecular weight, there is a maximum difference in solubility parameters that may be tolerated in order that the polymers be miscible. For most polymer pairs, the δ's are too greatly separated for miscibility, so that they lead to commingled mixtures. This would happen, for example, if polystyrene and poly(methyl methacrylate) or polycarbonate and poly(styrene-co-acrylonitrile) (SAN), or even two SAN's of differing AN content were mixed. There are cases where miscibility does occur. For example, high and low density polyethylenes are usually miscible, but immiscibility can arise if a linear polyethylene is mixed with a highly branched one.

Temperature dependence of Miscibility

In the above discussion, χ is predicted to be inversely proportional to T. With a positive χ, this leads to a decreasingly unfavorable enthalpic factor and decreasing miscibility with increasing temperature. This behavior, where solubility increases with increasing temperature is referred to as an *upper critical solution temperature* (UCST). In many cases the opposite phenomenon of a *lower critical solution temperature* (LCST) is found where miscibility becomes less as the temperature is increased. Also, a more complex variation of with T such as

$$\chi = A + B/T$$

is found. Deviations from the simple behavior arise from departure from the oversimplified assumptions involved in the Flory-Huggins derivation. Such deviations arise from:

1. Changes in volume that occur upon mixing,
2. specific interactions (hydrogen bonding, ionic) between components which may be temperature dependent, and
3. deviations arising from the failure of the lattice calculation to describe the way in which real molecules pack.

Improvements, both in elaborating the lattice description and in alternate approaches such as using radial distribution functions, have been carried out with some success.

Interfaces Between Immiscible Phases

The sharpness of the interface between two polymers also depends upon the χ for the pair which depends upon the difference between their δ values *(5)*. A large difference, such as might occur between polyethylene and poly(methyl methacrylate), the interface will be sharp leading to little interpenetration of the two polymers. However, as the δ's become closer, the interface becomes more diffuse (and usually stronger).

Thus, diffuse interfaces can be promoted if means are available for adjusting the δ's so as to bring them closer together. This is not easily accomplished, since δ depends upon the composition of the polymer itself. Means are through functionalization or blending. A miscible polymer may be added to an immiscible pair so as to shift the δ's of the blend, For example, adding poly(vinylidene fluoride) to an immiscible mixture of poly(methyl methacrylate) and poly(ethyl methacrylate), a miscible mixture may result. A more ready technique is to modify the local composition of polymer near a surface to as to strengthen its interface with a second polymer.

A related approach is to add a polymer which may act like an emulsifying agent or detergent to serve as a compatibilizer and bind the phases together, For example, a block copolymer with each block being miscible with one of the phases serves this role. The thermodynamics of mixing for such three component systems consisting of two homopolymers and a block copolymer have been worked out.*(6)* It is necessary, of course, that each block of the block copolymer be sufficiently long that it may be entangled with the homopolymer to impart sufficient mechanical strength to avoid failure at the interface.

Some Advantages of Immiscibility

Miscible mixtures often obey the "rule of mixtures" *(7)* where some property, such as the modulus, E, may be described by an equation of the sort

$$E = \phi_1 E_1 + \phi_2 E_2 + \phi_1 \phi_2 E_{12}$$

The E_{12} term represents some sort of interaction between components, and is zero if the "rule" strictly applies. Thus, properties may deviate positively or negatively from the linear variation predicted by the "rule" depending upon the sign of E_{12}. In any case, one usually finds behavior intermediate between that of the components.

With phase separated mixtures, the rule does not work. Most commonly, for non-interacting components, ultimate properties such as tensile strength, are often inferior for the mixture because of failure at the boundaries. If the loss in properties is not too bad, it is sometimes tolerated if an acceptable cheaper product can result by adding a cheap filler to a more expensive matrix polymer. For this purpose, mineral fillers are sometimes added to polymers.

If the immiscible phases interact, enhancement of properties may result. For example, adding carbon black to rubber (where there is strong interaction between the carbon black and the rubber) is a common practice to reinforce the rubber. A fibrous structure and small particle size are desirable to maximize the interaction.

The moduli of a filled polymer with dilute, rigid spherical filler particles may be described by

$$E = E_o [1 + 2.5\phi + 14.1\phi^2 + \dots]$$

E_o is the modulus of the unfilled polymer and ϕ is the volume fraction of filler. The equation may be generalized for other shaped particles and to cases where the particles are deformable. For oriented anisotropically shaped particles, the moduli become anisotropic. This may describe the moduli of oriented fiber-filled composites.

At higher concentrations of filler, interactions between particles become important, and properties lie between those predicted by a parallel (equal strain) model

$$E = \phi_1 E_1 + \phi_2 E_2$$

where moduli are additive to that of the series (equal stress) model

$$(1/E) = \phi_1 (1/E_1) + \phi_2 (1/E_2)$$

where compliences are additive. Takiyanagi (8) has proposed a "generalized series-parallel model with parameters describing the "degree of parallelness" dependent upon morphology. In such systems, one may find "phase inversion" phenomena. For example, if one adds a high modulus component to a lower modulus matrix, at a concentration where the high modulus particles begin to contact, the modulus may rapidly increase from a low to a high value. This then represents very non-linear behavior. Thus, there is a flexibility in property variation in phase separated systems where the way in which the properties of the components combine depends upon the morphology.

Fracture of a polymer usually occurs at a stress considerably lower than that estimated for a perfectly homogeneous polymer. This happens because of the presence " of defects or "cracks". Growth of such cracks occurs when the energy decrease associated with the loss of stored elastic energy exceeds the energy increase associated with the formation of new crack surface according to the "Griffith criterion" (9). The "toughness" of polymers is often greater than that for other materials because viscoelastic energy dissipation also gets included. With "impact improved" polymers, a rubbery phase is included within a brittle polymer which induces craze or shear band generation within the polymer and adds to the energy dissipation, thus imparting toughness. There is the potential that under proper conditions, such property improvement could be achieved with recycled phase separated polymers.

Crystallization of Mixed Polymers

Blends dealt with in recycling often involve components where one or both may crystallize from the melt. Such mixtures could be classified into those where:

1. The melts are immiscible and one component crystallizes.
2. The melts are immiscible and both components crystallize.
3. The melts are miscible and one component crystallizes
4. The melts are miscible and both components crystallize.

As an example of type (1), one might consider polypropylene mixed with an ethylene-propylene copolymer of such composition (50:50) that the copolymer is amorphous. (Such mixtures are sometimes made to impact improve the polypropylene.) In this case, of course, the crystals will be confined to the crystallizable phase so that their morphology will be limited by that of the phase prior to crystallization. The morphology and kinetics may be altered by the influence of the interface which could, for example, be active for nucleation and lead to the development of a trans-crystalline layer. Also, if crystallization is heterogeneous, the dispersed crystallizable phase could be sufficiently small so that it does not contain a nucleus and its crystallization could be retarded. The resulting structure will consist of crystalline and amorphous regions arranged in a manner determined by that of the melt prior to crystallization.

Crystallization of such mixtures often occurs during processing, so that the phases may be oriented as a consequence of shear. Then, crystallization may be affected by such orientation. The relative orientation of the two phases depends upon their melt viscosities. If, for example, the crystallizable phase is dispersed in a high viscosity amorphous matrix, it may experience appreciable orientation.

An example of a type (2) mixture would be the same pair where the copolymer was rich in polypropylene so that it also crystallizes. In this case, the polypropylene phase, having the higher melting point, crystallizes first. As a consequence of the volume changes occurring with its crystallization, strains may be imposed on the copolymer phase which could affect its crystallization. One would produce a composite structure containing two species of crystals with a morphology determined by that of the phase separated melt prior to crystallization.

A type (3) mixture is that of poly(vinylidene fluoride) (PVDF) with poly(methyl methacrylate) (PMMA). In this case, both components are normally miscible in the melt and the PVDF crystallizes *(10)*. The crystallization rate of the PVDF, being dependent upon the glass temperature of the melt, decreases with increasing concentration of the high Tg PMMA component.

The PMMA is excluded as crystallization proceeds and resides in the interlamellar region. It's presence there can be demonstrated by small-angle x-ray scattering (SAXS) *(11)* where one finds that:

1. The lamellar spacing increases with increasing amounts of PMMA, and
2. the scattering intensity increases with increasing amount of PMMA. This occurs because PMMA has a lower electron density than does PVDF, so its presence in the interlamellar region increases the electron density difference between the amorphous phase and the crystalline PVDF with a consequent increase in intensity.

A similar morphology is found for poly(caprolactone((PCL)/ poly(vinyl chloride)(PVC) blends *(12)*. However, in this case, since the electron density of the PVC is higher than that of the PCL, the exclusion of the mostly amorphous PVC increases the electron density of the amorphous phase and thus decreases the difference between it and the crystalline PCL with a consequent decrease in intensity.

The excluded amorphous component does not necessarily remain between the lamellae. For a blend of isotactic with atactic polystyrene (PS), the lamallar spacing of the crystalline isotactic PS is unaffected by the presence of the excluded atactic component indicating that it does not reside between the lamellae but diffuses into separate domains *(13)*. It is reasonable that it should, since the entropy of a PS molecule in the interlamellar region is lower that when it is not confined. Thus diffusion out of the interlamellae region involves an entropy increase and a free energy decrease. In fact, the reason why in the PVDF/PMMA and the PCL/PVC systems, the excluded component remains between the lamellae is that there is an attractive interaction between the components, as indicated by a negative χ, so the diffusion out of these regions incurs a positive ΔH_{mix} which offsets the positive ΔS_{mix} and renders the free energy change positive.

The amorphous phase between the lamellae may not be homogeneous. For the PVDF/PMMA system, there is evidence that the PVDF concentration in the amorphous region close to the PVDF crystals is higher than the average *(14)*. This may be a consequence of the PVDF chains in this region having reduced mobility because of their being anchored to the crystals. This would result in a lower entropy of mixing in this region.

At a larger morphological level, if the crystallizable component, forms spherulites, it is likely that the rejected amorphous component will remain within the spherulites (unless it can diffuse faster than the spherulite radial growth rate).

Thus, in this case, the resulting morphology will be that of a composite of crystalline and amorphous polymer arranged in a manner dependent upon the competition of crystallization and diffusion kinetics. These will depend upon composition, temperature, and the molecular weights of the components.

Case (4), where both components may crystallize from a homogeneous melt, may be represented by a mixture of high density (HDPE) and low density (LDPE) polyethylene. With such a mixture, one component (the HDPE) crystallizes first because of its higher melting point *(15)*. The other component will initially be excluded from the crystals into the interlamellae region from which it will tend to diffuse (because the χ will be close to zero). However, if the temperature is dropping during the course of the crystallization, the LDPE may crystallize before it escapes from the interlamellae region. Thus, morphology depends upon the temperature history.

Evidence for the above comes from SAXS studies. If diffusion of the LDPE has occurred, then crystallization of the HDPE and the LDPE will occur separately and SAXS spacings characteristic of each will be found. However, if crystallization occurs before diffusion, LDPE crystals will form between those of HDPE giving rise to intermediate SAXS spacings. Experimentally, both situations are found, depending upon the degree of quench of the sample *(16)*.

The fact that the LDPE crystallizes within the first formed spherulites of HDPE means that the spherulite size is determined by the HDPE component. This accounts for the observation (by small-angle light scattering *(17)*) of the variation of spherulite size with composition of a HDPE/LDPE blend. Spherulites of LDPE are normally smaller than those of HDPE. However, the addition of rather small amounts (5%) of HDPE to a LDPE melt results in an increase in spherulite size to a value approaching that for HDPE.

In this case, crystallization kinetics are affected by epitaxial crystallization. Adding rather small amounts of HDPE to LDPE appreciably increases its rate of crystallization. This occurs because the LDPE crystals are nucleated by the previously grown crystals of HDPE *(17)*.

Co-crystallization must also be considered. Branches larger than methyl are normally excluded from polyethylene crystals. This accounts for the observation that crystal thicknesses (and melting points) of branched polyethylenes are generally lower than those for linear polyethylenes. In crystallization of a HDPE/LDPE mixture, unbranched sequences of the LDPE may co-crystallize with the linear polyethylene. This will occur at high degrees of quench when the two components crystallize simultaneously. At low degrees of quench where the crystallization of the HDPE can be almost completed prior to that of the LDPE, less co-crystallization occurs. Co-crystallization is more prevalent with blends of HDPE with linear low density polyethylenes (LLDPE) having low degrees of branching and melting points closer to that of the HDPE. The occurrence of co-crystallization can be demonstrated using differential scanning calorimetry (DSC) or by observing splitting of infrared bands in blends where one of the components is deuterium substituted *(18)*.

With high degrees of branching (10-20 branches per hundred CH_2's, depending on molecular weight branch length), phase separation of the amorphous melt may occur, so considerations of Case (3) would then apply.

It is evident that a variety of morphologies may ensue in a manner dependent upon the choice of components and crystallization conditions. The mechanical, optical, transport, and other properties are morphology dependent. As an example, the clarity of LDPE is substantially reduced upon adding small amounts of HDPE *(19)*. This is primarily a result if the increase in spherulite size (since scattering depends upon R^3). Secondary effects arise from changes in the internal morphology of the spherulites.

Conclusions

Recycling of polymers requires dealing with polymer mixtures. Most of these will be immiscible and will lead to products having lowered physical properties associated with weak interfaces between components. Miscibility principles can guide means for forming diffuse or stronger interfaces with enhanced properties.

Various means are available to increase the miscibility of polymer mixtures. Even with miscible mixtures, interactions between components may be important. This is especially true if one or both components crystallize, since the presence of a second component generally affects the crystallization of the first.

These concepts may even be important for separated polymers, where the components mixed together are the same chemical species. In this case, they are likely to differ in molecular weight (and/or distribution), branching, tacticity, etc. so differences arise from the consequences of mixing such species. The result will be morphologies dependent upon the nature of the species and upon the processing conditions.

From a knowledge of the principles governing morphologies and the consequent properties of such mixtures, intelligent choices of processing conditions and applications may be possible. For example, property improvement might be achieved by adding a virgin polymer to a recycling feedstock so as to modify its morphology in a desirable way.

Acknowledgement
The author appreciates the support of the Materials Research Laboratory of the University of Massachusetts, the National Science Foundation Division of Materials Research, and the Exxon Chemical Company. Much of the work was accomplished in collaboration with several graduate students, post-doctoral fellows, and visiting scientists whose names are indicated in the appropriate references. The granting of beam time and technical assistance with SANS at the national facilities of Oak Ridge and Argonne National Laboratories and the National Institute for Standards and Technology (NIST) was invaluable.

Literature Cited

1. *Plastics & the Solid Waste Issue,* Exxon Chemical Company, 1991
2. Flory, P. J., *Principles of Polymer Chemistry,* Cornell University Press, Ithaca, NY, 1953
3. Hildeband, J. and Scott, R., *Solubility of Non-Electrolytes,* 3 Edition, Dover, New York
4. London, F, Z. *Physik*, 1930, *63*. p. 245
5. Helfand, E, *J. Chem. Phys.*, 1975, *62*, p. 999
6. Noolandi and Hong, K. M., *Macromolecules*, 1982, *15* , 482; 1983, *16*, 1983
7. Nielsen, L., *Mechanical Properties of Polymers*, Reinhold Publishing Corp., New York, NY, 1962, Chapt.6
8. Takayanagi, M., xxx
9. Griffith, A. A., *Philos. Trans. Roy. Soc. London* , 1920, *221A*, p. 163
10. Morra, B.S, *J. Polym. Sci., Polym. Phys. Ed.*, 1982, *20*, pp. 2243-2259
11. Morra, B.S., *Studies of Melt Crystallized Poly(viylidene Fluoride,/ Poly(methyl methacrylate) Blends* , PhD. Dissertation, University of Massachusetts, Amherst, MA, 1980
12. Russell, T.P. and Stein, R.S., *J. Polym. Sci., Polym. Phys. Ed.*, 1983, *21*, pp. 999-1010
13. Wei, M. PhD. Dissertation, University of Massachusetts, Amherst, MA,
14. Hermans, W. PhD. Dissertation, University of Massachusetts, Amherst, MA,
15. Hu, S-R, Kyu, and Stein, R.S., J. *Polym. Sci., Polym. Phys. Ed.*, 1987, *25*, pp. 71-87
16. Ree, M., PhD. Dissertation, University of Massachusetts, Amherst, MA,
17. McGuire, S. *Scattering Studies of Solid Polymers: Polyethylene Blends.* M.S. Dissertation, University of Massachusetts, Amherst, MA, 1933
18. Tashiro, K., Stein, R.S., and Hsu, S.L., *Macromolecules,* in press
19. Ree, M., Kyu, T., and Stein, R.S., *J. Polym. Sci., Polym. Phys. Ed.*, 1987, *25*, pp. 105-126

RECEIVED August 3, 1992

Chapter 5

Automotive Recycling in Japan

Kaoru Asakawa

Materials Research Laboratory, Nissan Research Center, Nissan Motor
Company, Ltd., 1, Natsushima-cho, Yokosuka 237, Japan

In Japan, the lack of sufficient landfill capacity is again generating a great deal of public interest in the question of waste disposal.   Scrapped car bodies in Japan are shredded into small bits and pieces, referred to as shredder dust. The cost of disposing of that dust has skyrocketed recently,causing a problem for the automotive industry.
In some areas of the country, old vehicles abandoned on the road have become a fairly common sight.

These circumstances set stage for Nissan's decision to establish a Recycling Promotion Committee in August of 1990.   Up to that point, each department at Nissan had carried out its own recycling effort.   The reason for establishing the committee was to pull together all of those individual efforts, gather the latest information and, on that basis, try to find broad-based solutions to the problem of waste disposal.

This paper focuses on the issue of how to recycle plastic waste, and will explain our thinking on this issue in terms of the technology involved and describe some of the efforts under way at Nissan to deal with this problem.

Current Situation on Scrapping Vehicle in Japan

Fig.1 shows the total amount of waste generated in Japan in 1987. Approximately 70% of the ordinary waste in Japan is incinerated and then buried in landfills. The growing volume of waste in recent years has led to a shortage of incinerator capacity. As a result, some waste is now disposed of at landfills directly without being incinerated. Industrial waste accounts for more than four times the volume of ordinary waste. The waste disposal situation is growing more severe all the time.

Fig.2   shows the decreasing capacity of landfills in the Tokyo metropolitan area. As you can see, the landfill capacity has been reduced by one-half in the last three years. It is becoming increasingly more difficult to find waste disposal sites in this area. As a result, some of the disposers of industrial waste are going as far as 500 kilometers away to find landfill sites and are transporting waste there for disposal.
This is one reason for the soaring cost of waste disposal and it is also giving rise to various social problems.

Against this social backdrop, what is the situation like for cars?
Fig.3   shows the number of new car registrations and the number of scrapped cars during the past decade.   Two things that stand out here are the sharp rise in new car registrations from around 1986 and the accompanying increase in scrapped vehicles.

NOTE: This chapter is based on a paper presented at the 24th ISATA symposium, Florence, May 1991. ISATA: 42 Lloyd Park Avenue Croydon CRO 5SB, United Kingdom

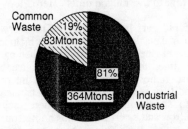

Figure 1. Total Volume of Waste in Japan

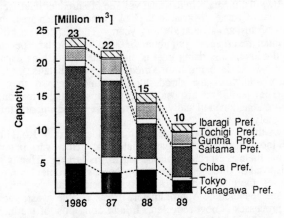

Figure 2. Capacity of Landfills in Tokoyo Metropolitan Area

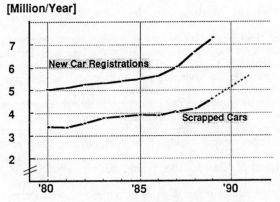

Figure 3. Number of Scrapped Cars in Japan

In 1989, some 4,620,000 cars were scrapped. The capacity for scrapping old cars is estimated at between four to five million units. Here, too, the situation is continually becoming more severe.

The scrapping of old cars begins with the recovery of valuable items, which disassemblers resell to make a living. These items include catalytic converters, parts that are still usable, and engine components and other parts with a high aluminum content. At the same time, the tires, battery, gasoline tank and other potentially dangerous parts are also removed, as they would interfere with the subsequent scrapping operation, and they are processed by a separate route. The stripped down bodies are then sent to shredding companies where they are chopped to bits by shredding machines. Iron and steel and major nonferrous metals are then separated and reclaimed. These metals are a source of income for the shredding companies which sell them to other processors. However, in many cases the shredder dust that remains can only be disposed of as landfill. (Fig.4)

Fig.5 shows the increase in the total volume of shredder dust generated in Japan and the rising cost of waste disposal in the Tokyo metropolitan area. Nowadays, over one million tons of dust are generated every year. The disposal cost of that dust in the Tokyo metropolitan area ranges from about $100 to $160 dollars per ton.

As I mentioned above, this rising cost is due to the diminishing capacity of landfills and to fact that waste is being transported such long distances. This has greatly affected the profitability of the shredding companies and so they are asking the disassemblers to pay part of the disposal cost. Because of this situation, the number of old vehicles being abandoned illegally on the road is increasing and this has become a social problem lately.

Fig.6 shows the material components of shredder dust. Roughly, the dust is a mixture of 50% inorganic and 50% organic materials. However, in terms of volume, the percentage of them, especially plastic waste, is larger. This is one of the main reasons why disposal of shredder dust has now become a problem.

Fig.7 shows the total amount of waste generated at Nissan and the various amounts that are recycled. Approximately 75% of the waste generated in our manufacturing processes is now recycled. In addition, most of the incinerators at our plants have been outfitted with advanced anti-pollution devices. Through incineration, plant waste is recycled into energy, which is used again in the vehicle manufacturing process. We also have our own landfill sites.

NISSAN's Current Actions on Recycling

In August of 1990, Nissan became the first vehicle manufacturer in Japan to set up a Recycling Promotion Committee. This committee was formed to provide a company-wide organization that could promote recycling more effectively. The committee is responsible for gathering information and for implementing activities which in the past were carried out by individual departments. Fig.8 shows the organization of this committee. The committee is actively engaged in many different activities that go beyond just technical development efforts. Working groups have been formed to examine the infrastructure needed for recycling and to promote cooperation with organizations outside of the company, including government bodies and the Japan Automobile Manufacturers Association. In addition, we have also set up subcommittees in Europe and North America with the aim of carrying out global recycling activities.

Fig.9 shows an outline of our thinking on recycling from a technical standpoint. As I mentioned above, one of the most pressing problems right now is the recycling of plastics, and that is why we decided to take up this issue first.

Fig.10 shows specific examples of plastic parts and materials that Nissan is now recycling. Many of these parts and materials are being recycled by our suppliers. We manufacture plastic bumpers and fuel tanks in-house and recycle the scrap

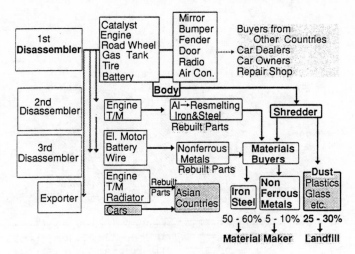

Figure 4. Car Scrapping Process in Japan

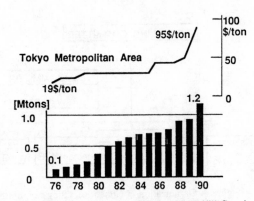

Figure 5. Total Volume of Shredder Dust and Landfill Cost in Japan

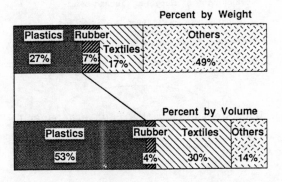

Figure 6. Material Components of Shredder Dust

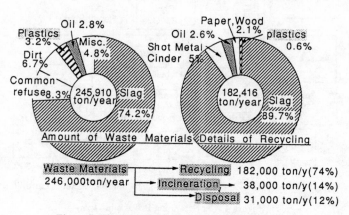

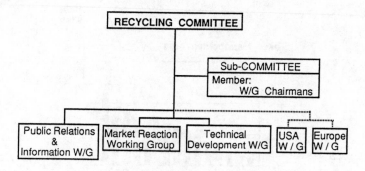

Figure 7. Waste Materials and Their Recycling at Nissan

Figure 8. Organization of Nissan Recycling Committee

Development of Recycling Technologies

1. Harmony between customer satisfaction and preservation of the natural environment.
2. Consideration of vehicle life from design to scrappage based on a global perspective.
3. Effective use of resources and environmental protection.

Cars are born from the Earth and return to it

Figure 9. Nissan's Thinking on Recycling Technologies

materials that are generated in the manufacturing process for these parts.   Plastic materials recycled in other industries are also being used in manufacturing new vehicles. We also mark certain plastic parts with an in-house code to identify the type of material used.

Fig.11 shows some of the major technical issues that we are working on right now in connection with the recycling of plastics.   Here, this paper focuses on some specific examples with respect to materials engineering.

Fig.12 shows the growing use of plastics in Nissan models and in all Japanese vehicles in general.   These figures indicate the percentage of the vehicle weight accounted for by plastics.   Up to now, plastic parts have been very useful because they can provide weight reductions, assurance of safety, comfort enhancement and greater design freedom -- all at a reasonable cost.   However, today there is a need to apply a new evaluation standard that takes recyclability into account.   This standard should be applied not only in evaluating different types of plastics but also in making comparisons between plastics and other types of materials.   Depending on the results of such comparisons and on whether the technologies are developed for using plastics to the best advantage, the amount of plastics used in vehicles in the future will vary considerably.

One of our priority issues at Nissan is to switch to types of plastic that are easier to recycle and are more compatible with environmental concerns.   The chart at the left in Fig.13 shows a breakdown of the various types of plastic that we used in 1988.   The one at the right shows a similar breakdown for the plastics used in the Nissan "PRIMERA(INFINITI G20)" which we released in 1990.   These figures are indicative of the positive effort we are making to switch to polypropylene, polyethylene and other types of plastic that are easier to recycle and are friendlier to the environment.   We intend to continue this policy in the future and vigorous efforts are now under way to apply these environmentally friendly plastics to more advanced components.

Among the plastic parts now in use, polypropylene bumpers have a relatively simply material composition.   And also because of their large size, they are being targeted for recycling.   Nissan first applied polypropylene to bumpers at an early date and we have built up ample technical capabilities in this area.

These are the recent major results of our R&D activities concerning the technology for recycling polypropylene

(1)We found that the mechanical properties and qualities of recycled PP are lowered more by the flagments of top coat (paint) than by the material deterioration or any other reasons.

(2)We have succeeded in removing the top coat from bumpers whith an aqueous solution of organic salt (Aqueous Paint Decomposer), which is harmless to humans and the environment (see Fig.14).

(3)The recycled PP without the top coat, retains more than 95% of its mechanical properties when compared to the virgin one (see Fig.15).

(4)The recycled PP has the same moldability as the original.

(5)The molded PP bumper, made from 100% recycled PP, is almost equal to the original bumper from the standpoint of function and quality.

(*Fig.14,15 are presented by Nissan at "GLOBE'92", March 19 in Vancouver*)

Once the recycling of bumpers begins in earnest, various grades of materials from different model years will all be collected together.   If recycling is to succeed, it will be necessary to identify variations in quality accurately.   In order to gather basic data for this purpose, we are now conducting an experiment in which we are recovering and examining old cars from the market on a systematic basis.   At the same time, we are also taking advantage of this experiment to investigate ease of disassembly and possibilities for recycling other plastic parts.

Recycling in Manufacturing
- Parts
  - Washer Tank, Wiper Parts, Rear Combination Lamp, Heater Case
    Cooler Case, Blower Case, PP Bumper, Rad. Reserver Tank
- Materials
  - PP, POM, ABS, AAS, PMMA, ASPE

Application of Recycled Plastics
- Parts
  - Fusible Insulator, Floor Insulator, Cushon Pad, Door Trim Base
    Floor Mat, Trunk Trim, Engine Under Cover
- Materials
  - Chip PUR, PE Sheet(Agricultural Use), Waste Textiles, Waste PP
    Waste PVC, Kiln ashes(Paint waste)

Identification of Plastic Materials by Marking
- Parts
  - Bumper, Rad. Grille, Fender Protector, Engine Under Cover
    Heater Case, Cooler Case, Blower Case, A/C Duct
- Materials
  - PP, PE, PUR, ABS, PC

Figure 10. Examples of Plastic Recycling at Nissan

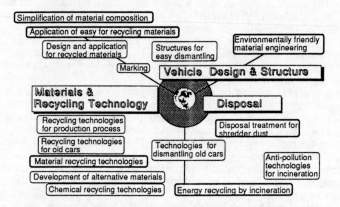

Figure 11. Technical Issues Involved in Plastic Recycling

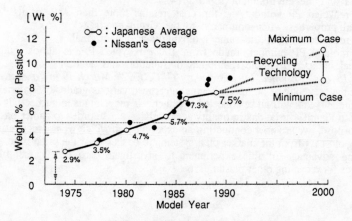

Figure 12. Trends of Plastics Application at Nissan

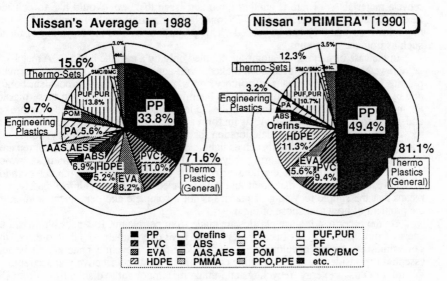

Figure 13. Proportions of Different Plastics in Nissan's Cars

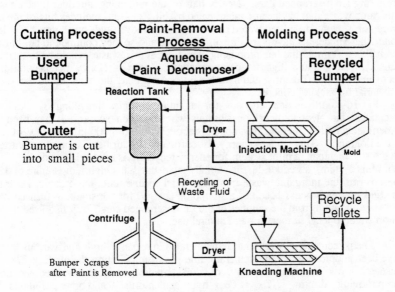

Figure 14. Polypropylene Bumper Recycling Process

Material Recycling & Energy Recycling

So far I have been explaining   about some of our recent activities at Nissan t
recycle materials.   There is another important issue that we   would like to conside
next and that is energy recycling.   Thermoplastic materials can be recycled by meltin
them down so that they can be reused. This process resembles the recycling of metal
such as iron and aluminum.

In the case of metals, the atoms are coupled only by a metallic bond.   A metal ca
be renewed because after it has been melted it will again form a solid. However, man
plastics form a solid by means of primary bonds, consisting mainly of covalent bonds
and by secondary bonds held together by intermolecular force.   When these plastic
are melted down during recycling,   only the low-energy secondary bonds are broke
and upon cooling the material will again form a solid.   During their service life prior t
recycling,   plastics undergo deterioration because some of the primary bonds ar
damaged by exposure to stimuli such as light and heat.   That deterioration is not repaire
when they are melted again during recycling. Consequently, unlike metals, which
theoretically can be renewed any number of times, plastics accumulate the deterioratior
they have suffered up to that point and it is recycled with them. Unfortunately, thi
means that the simple recycling of plastics is not a complete and permanent solution t
the problem of how to dispose of them.

On the other hand, since plastics are made from petroleum, it is possible to extrac
energy from them. This is one advantage they have over metals.   Therefore, at the
same time that we are developing   technologies for   recycling materials, it is als
essential to consider ways of extracting energy from plastics effectively and safely.
We refer to this as energy recycling and I think it is important to distinguish it from the
simple incineration of waste.

As mentioned in Fig.6,   plastics account for around 27% of the shredder dus
resulting from scrapped cars.   Nearly half of the remaining shredder dust consists o
inorganic waste that has been contaminated by oil.   This includes bits of metal tha
were not separated and recovered earlier, glass, sludge and other items.   In tha
state, these materials have very little usable value and so they continue to be buried in
landfills.   I think that one promising way to treat this waste would be to turn it into
slag by heating it at a high temperature above 1200°C.   Heavy metals could then be
recovered more efficiently and harmlessly.   After that, the slag could be used as
a material for making concrete.

Fig.16 outlines one concept for a future plastics recycling system which
integrates the technological possibilities we have seen so far.   It has been found
experimentally that shredder dust contains about 3,500 to 4,000 kilocalories of energy
per kilogram.   That much energy is sufficient to turn the inorganic waste into
slag.   In Japan, a pilot system for energy recycling is now being tested.
With this system, shredder dust is partially incinerated at a high temperature of 1300°C
to transform the inorganic waste into slag. At the same time, the organic waste that is
gasified by this process becomes a source of energy for the smelting furnace used to
recover aluminum from the engines of the same scrapped cars. It is also used to supply
warm water to local farmers for agricultural use.

The percentage of waste that is incinerated in Japan is high and various measures
have been adopted to prevent incinerator flue gas from polluting the air (Fig.17).
Research work is still under way, of course, to develop more advanced technology
for removing dioxin, NOx, SOx, hydrogen chloride   and other pollutants from
incinerator emissions.   This technology will have to be incorporated into a practical
system that can be made available at a lower cost.

Our ideas on recycling are incorporated in this future system for recycling
automotive plastics.   As much as possible, we want to recycle materials and energy
and turn waste into usable resources.   We hope that this can be done at a level that

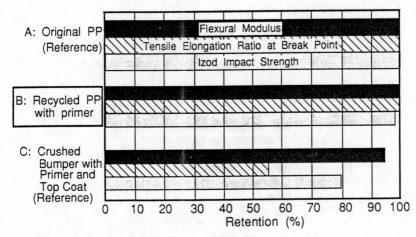

Figure 15. Mechanical Properties (at R.T.) of Old Bumper's PP (Elapsed Time: 3 to 5 Years)

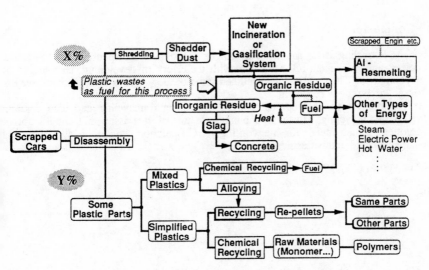

Figure 16. Future Recycling System for Automotive Plastics

is cost effective, achieves an optimum energy balance, provides reliable protectio
of   the environment and can be accomplished within the framework of existing soci
systems.   We plan to move ahead with our research activities aimed at achieving th
goal.

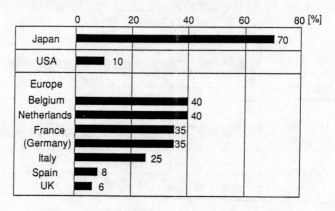

Figure 17. Rate of Waste Disposal by Incineration.

Conclusion

Recycling is an activity that will be absolutely indispensable to tomorrow'
society, and we are in full agreement with the effort of developing tecnologies fo
recycling.   This paper has outlined some of the vigorous efforts under way at
Nissan to promote recycling.

We continue to develop technologies for material recycling and also research
for energy recycling. However, one major feature of this issue that should not be
forgotten is the need for networking.   Cooperation involving all   types of
businesses and industries is essential.

RECEIVED April 17, 1992

Chapter 6

Pyrolysis of Polymers

Walter Kaminsky

Institute for Technical and Macromolecular Chemistry, University
of Hamburg, Bundesstrasse 45, D–2000 Hamburg 13, Germany

Oil and gas may be retrieved by the pyrolysis of plastics and rubber wastes. Processes have been carried out in melting
vessels, rotary kilns and fluidized beds of
sand. In some cases i.e., polystyrene or
polymethylmethacrylate (PMMA) it is possible
to win back the monomers by pyrolysis.
Fluidized bed pyrolysis has turned out to
be particularly advantageous because of the
high heat and mass transfer. Under the
appropriate conditions these processes
could be successful in the market.

The recycling of plastic and rubber wastes is of increasing importance as landfilling and incineration becomes more expensive and the acceptance of these methods is decreasing. Most plastics are produced from oil
and can be pyrolyzed into petrochemicals. As plastic
waste consists mainly of polyolefins, the final product
would be rich in olefins. Mixed, contaminated, and combined plastics can be pyrolyzed, too.

Pyrolysis is complicated by the fact that plastics,
rubbers, or biopolymers are poor thermal conductors,
whereas the degradation of macromolecules requires considerable amounts of energy.

Processes

The pyrolysis of plastic wastes, used tires, and biopolymers has been studied in melting vessels, blast
furnaces, autoclaves, tube reactors, rotary kilns,
cooking chambers, and fluidized bed reactors (1-7).

0097–6156/92/0513–0060$06.00/0
© 1992 American Chemical Society

In the course of the development rotary kilns and
fluidized beds have turned out to be the most suitable
pyrolysis aggregates (*8-15*). Table 1 compiles various
pyrolysis processes for plastic wastes, rubber, and scrap
tires that have already reached the stage of industrial
testing.

Meltable plastic wastes can also be decomposed in
agitator vessels. The vessels can be heated externally or
by means of internal piping. A combination of initial de-
hydrochlorination and melting followed by thermal degradation was developed by Mitsubishi Heavy Industries (*16*).
Other processes are operating in autoclaves for the hydrogenation of plastics at 300 °C and 200-400 bar hydrogen pressure. Rotary kiln processes are particularly numerous. They are marked by relatively long residence
times of the waste in the reactor of 20 minutes and more,
whereas dwell times in fluidized bed reactors hardly exceed a few seconds with a 1,5 minute maximum.

One of the first processes was the Kobe Steel process (*17*) using a rotary kiln for the pyrolysis of scrap
tires. The process ran over a long period and was combined with a cement factory. The pyrolysis products were
used instead of heating oil.

Tosco (*18*) used a rotary kiln for the pyrolysis of
scrap tires. The main interest was to recover carbon
black which could be reused by the rubber industry.

The core of the Salzgitter pyrolysis (Table 1) is an
externally heated rotary kiln. This 26 m long reactor
with a diameter of 2,8 m is subdivided into six segments
that can be heated separately. The pyrolysis takes place
in the reactor at temperatures of 400 - 700 °C and yields
pyrolysis gas and oil as well as solid residues. The purified pyrolysis gas is used as fuel for operating the
plant and excess gas is converted to electricity in a
linked power plant.

A fluidized bed has a number of special advantages
for the pyrolysis because it is characterized by excellent heat and mass transfer as well as constant temperature throughout the reactor. The fluidized bed is
generated by a flow of air or an inert gas that is directed from below through a layer of fine grained material,
e.g. sand or carbon black, at a flow rate that is sufficient to create a swirl in the bed. At this stage the
fluidized bed behaves like a liquid.

Table 1. Processes for Pyrolysis of Plastic Wastes.

Name of Process	Process	Products	Size, Site
DBA-Process (BKMI)	Rotary kiln, indirect heated, 450–500 °C	Energy	6 t/h, Günzburg/Germany
Ebara-Process	Two fluidized beds	Energy	4 t/h, Yokohama/Japan
Kobe-Steel-Process	Rotary kiln, indirect heated, 500–700 °C	Oil, gas, energy	1 t/h, Kobe/Japan
Tosco II-Process	Rotary kiln, indirect heated, 500–550 °C	Energy, carbon black	Pilot plant, Golden/USA
KWU-Process	Rotary kiln, indirect heated, 450–500 °C	Energy	3 t/h, Ulm/Germany
Dr. Otto Noell-Process	Rotary kiln, indirect heated, 650–700 °C	Oil, gas	6 t/h, Salzgitter/Germany
Mitsubishi Heavy Industries	Melting vessel, indirect heated, 550 °C	Oil	100 kg/h, Japan
Energas-Process	Rotary kiln, indirect heated, 650–750 °C	Energy	150 kg/h, Gladbeck/Germany
Tsukishama Kikai-Process	Two fluidized beds, one oxidizing	Energy	3 x 6,25 t/h, Funabashi/Japan
Hamburg-Process	Fluidized bed, indirect heated, 600–900 °C	Oil, gas, carbon	20–60 kg/h, University of Hamburg; 0,5 t/h, Ebenhausen; 1 t/h, Grimma; all in Germany

In Japan, fluidized bed pyrolysis plants are operated with air or oxygen feed (*9*). The partial oxidation generates a part of the necessary fission energy while on the other hand a part of the products is burnt. The product oils are partly oxidized and their energy content is some 10 % below that of pure hydrocarbons.

Recent work at the University of Hamburg has focused on the suitability of plastic wastes, used tires, and waste oil residues as a source of olefins and other hydrocarbons after pyrolysis in a fluidized bed reactor. Laboratory plants with capacities of 60 - 3000 g/h and two pilot plants capable of processing 10 - 40 kg/h of plastic and 120 kg/h of used tires were installed (*6, 19-22*).

The core of the pilot plant (Figure 1) is a fluidized bed reactor with an inside diameter of 450 mm. An auxiliary fluidized bed of quartz sand with a temperature of 600 to 900 °C is used for the pyrolysis of plastics that are fed into the reactor through a double flap gate or screw. Pyrolysis gas preheated to 400 °C is used to create the swirl in the fluidized bed. The heat input takes place indirectly through radiative fire pipes that are heated by pyrolysis gas.

The exhaust gases of the burners are then directed through a heat exchanger. The product gases emerging from the fluidized bed are separated from residual carbon and fine dust in a cyclone and then cooled to room temperature by quenching with xylene product oil in a condenser. The gas flow is then directed through two packed condensation columns.

Condensed oil fractions are distilled in two columns using the fraction boiling from 135 - 145 °C (mostly xylenes) as a quenching medium. Also produced are tar with a high boiling range as well as two fractions, one rich in toluene and the other in benzene.

The gas, largely stripped of liquid products, passes to an electrostatic precipitator where it is freed from small droplets. Subsequently, it is compressed in five membrane compressors, connected in parallel, and stored in three gas tanks, each 0,5 m^3 in volume. Part of the gas serves as fuel for the heat radiation pipes; the remainder, preheated by gases in the heat exchanger, is used for fluidizing the sand bed. The excess gas is flared.

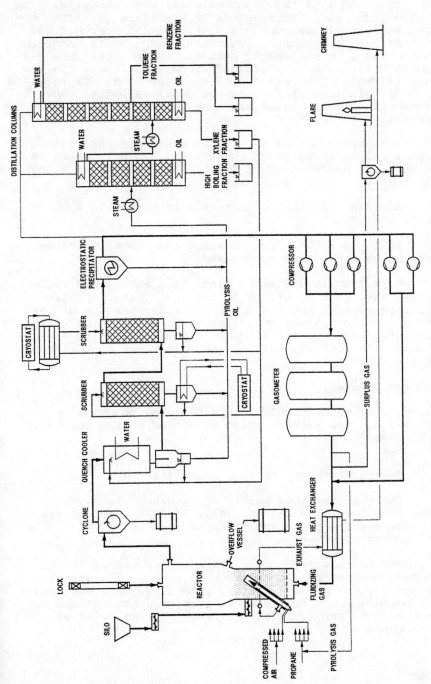

Figure 1. Scheme of the Technical Plant for Pyrolysis of Plastics

Up to 50 % of the input material may be retrieved in liquid form, which corresponds to a mixture of light benzene and bituminous coal tar with about 95 % aromatics. The oil may be processed into chemical products according to the usual petrochemical methods. Optimal reaction management aims at high yields of aromatics. Comparable raw material value analysis showed that the chemical processing of this type of high aromatic oil is better converted to valuable raw materials than used for heating or propulsion purposes.

The product gases are highly energetic fuel gases with a calorific value of about 50 MJ/m^3. Only about 40 % of them are needed as fuel for the heat radiation pipes, the rest is available as excess gas.

Any kind of plastic waste can be pyrolyzed even if it is soiled.

In Table 2 the input plastics and rubber which have been investigated are summarized.

More than 20 different kinds of plastics and rubber wastes have been pyrolyzed in the past years. Polyolefins, polyesters and rubber give similar amounts of gas and oil while polystyrene, PMMA, PVC, and lignine yields more oil than gas. All kinds of fillers like carbon black, metal oxides, silica, and metals are concentrated in the residue.

Monomer Recovery

Normally it is difficult to win back the monomers by pyrolysis. Pyrolysis of polyolefins led to the following fractions (wt.%):

Gas	20-90
Oil	5-45
Carbon black and filling materials	0.2-10

The gas consisted mainly of methane, ethane, ethylene, propylene, and butene (Table 3). The combined yield of ethylene and propylene produced from polyolefins does not exceed 60 wt.-%.

The concentration of olefins is influenced by the process parameters. If nitrogen is used as fluidizing medium for the pyrolysis of polyethylene, more ethylene and butadiene are formed as if pyrolysis gas is used as medium. Even pyrolysis of polypropylene led to 35 wt.-% of propylene produced.

Table 2. Pyrolysis of Plastic Waste in a Fluidized Bed. Different Feeds
and their Products

Feed	Pyrolysis (Temp. °C)	Gas (wt.-%)	Oil (wt.-%)	Residue (wt.-%)	Others (wt.-%)
Polyethylene PE	760	55,8	42,4	1,8 C	
Polypropylene PP	740	49,6	48,8	1,6 C	
Polystyrene PS	580	9,9	24,6	0,6	64,9 Styrene
Mixture PE/PP/PS	750	52,0	46,6	1,4	
Polyester	768	50,8	40,0	7,1	2,1 H_2O
Polyurethane	760	37,9	56,3	0,5	5,0 H_2O; 0,3 HCN
Polyamide PA-G	760	39,2	56,8	0,6	3,4 HCN
Polycarbonate	710	26,5	46,4	24,6	2,5 H_2O
Poly(methyl methacrylate)	450	1,25	1,4	0,15 C	97,2 MMA
Poly(vinyl chloride)*	740	6,8	28,1	8,8	56,3 HCl
Poly(tetrafluoroethylene)	760	89,3	10,4	0,3	
Medical syringes	720	56,3	36,4	5,8	1,5 Steel
Plastic from household waste separation	787	43,6	26,4	25,4	4,6 H_2O
Plastic from car-shredding	733	29,9	26,7	27.6	14,0 Metals, 1,8 H_2O
EPDM-Rubber**	700	32,3	19,2	47,5	1,0 H_2O
SB-Rubber***	740	25,1	31,9	42,8	0,2 H_2S
Scrap tires	700	22,4	27,1	39,0	11,5 Steel
Lignin	500	3,4	29,9	49,3	17,4 H_2O
Cellulose (Wood)	700	47,1	23,0	18,6 C	11,3 H_2O
Sewage Sludge	600	34,3	27.7	33,2	4,8 H_2O

* PVC: Poly(vinyl chloride)
** EPDM: Ethene-Propene-Diene-Monomers
*** SB: Styrene-Butadiene

Higher yields of monomers can be reached when polystyrene or polymethylmethacrylate (PMMA) are used. About 70 % of the monomeric styrene could be recovered by pyro-lysis. The liquid has to be purified in an expensive pro-cess to give polymerization-grade monomer. In contrast to this the fluidized bed pyrolysis of PMMA gave very large amounts (97 %) of monomeric methyl methacrylate (*23, 24*).

Rising the temperature increases the amount of gas. At temperatures beyond 550 °C the gas fraction increases up to 42 wt.-% at 590 °C. The gas consists of methane, ethene, propene, carbon monoxide and carbon dioxide.

Only small amounts of carbon black are formed. The main component in the liquid is methylmethacrylate (MMA). At a pyrolysis temperature of 450 °C it is 98,6 %, and at 490 °C 98,3 % pure. Even filled and coloured waste PMMA materials like rear lights gave a really clean monomer with 98,6 wt.-% MMA. The liquid contains small amounts of methyl isobutyrate, methyl acrylate, 1,4-cyclohexane dicarboxylic acid dimethyl ester (dimeric MMA), and methanol as side products. These concentrations are so low that the liquid could be polymerized to new PMMA after distillation without any further purification.

Petrochemicals

High yields of oil are obtained by pyrolysis in melting vessels (*4*). Polyolefins and polystyrene are particularly suitable feedstocks for this process.

The oil can account for up to 97 % of the feed. It is characterized by a high percentage of unsaturated end groups and can therefore be used for heating purposes only. The heating value ranges from 10 900 to 11 700 kcal/kg.

The content of unsaturates in the oils can be reduced by hydrogenation, thereby improving their storage stability. The high yields are hardly affected by this step.

Due to the large temperature gradient inside the rotary kiln, the product spectrum is very diverse. For this reason the gases and oils generated in the process are normally used for the direct generation of energy (heat).

In the fluidized bed process the liquids consist mainly of aromatics. Table 4 shows the composition of the pyrolysis products of polyethylene, plastics seperated

Table 3. Composition of Gases in the Pyrolysis of Polyethylene at
          Different Temperatures in wt.% and vol.%

Temperature °C	650	740	820	820
Compound	wt.%	wt.%	wt.%	vol.%
Hydrogen	0,61	1,26	2,57	21,71
Methane	18,52	39,10	50,27	53,09
Ethylene	21,09	32,85	32,45	19,58
Ethane	12,12	11,13	5,42	3,05
Propylene	24,27	9,20	2,10	0,85
Propane	2,52	0,39	0,11	0,04
Butene	10,21	0,89	0,08	0,03
Butadiene	1,30	0,42	1,40	0,44
Pentene	2,32	0,01	0,03	0,01
Cyclopentadiene	1,80	0,49	0,34	0,09
Other compounds	5,24	4,26	5,23	1,11
Density (g/cm^3)	1,161	0,922	0,704	0,704

Table 4. Pyrolysis Products from Different Kind of Plastic Wastes in a
          Fluidized Bed Process

Inputmaterials:	Polyethylene	Plastic waste	Plastic from car shredding	Scrap tires
Pyrolysis Temperature:	780 °C	790 °C	700 °C	720 °C
GAS	51,2	43,7	26,8	21,8
From this:				
Methane	19,8	17,5	6,1	8,4
Ethylene	19,7	11,4	3,2	1,8
Ethane	4,3	3,9	1,9	2,4
Propene	4,4	1,8	1,9	0,4
Carbon dioxide	-	2,0	4,1	1,5
Carbon monoxide	-	3,9	7,2	1,5
Hydrogen sulfide	-	-	-	0,01
OIL	46,2	26,3	21,4	21,0
From this:				
Benzene	23,9	12,4	4,1	2,6
Toluene	5,9	3,8	4,4	2,6
Naphthalene	3,7	2,4	0,9	0,5
Benzonitrile	-	-	0,8	-
Tar	7,2	4,0	6,2	11,5
Water	-	4,6	1,5	8,0
Soot, fillers	1,8	25,4	50,3	35,5
Steel	-	-	-	13,7

scrap tires. The collected plastic fraction seperated
from household waste contains 57 % polyolefins, 14 %
polyvinylchloride (PVC), 19 % polystyrene, 5 % other
plastics and paper, and 5 % inorganic materials (sand and
salts). The main compounds in the gas are methane,
ethane, ethylene, and propene; in the oil the main
compounds are benzene, toluene, and naphthalene. Carbon
dioxide and carbon monoxide are formed from polyester,
polyurethanes, polyamides, and cellulose-containing
materials. In the case of PVC containing plastics the
products contain a lot of hydrogen chloride and carbon
black.

By addition of lime to the fluidized bed, the hydrogen chloride from the decomposition of PVC can be bound
chemically. The calcium chloride that is formed in the
process, however, has a tendency to cloq the fluidized
bed at higher concentrations. Therefore it is attempted
to eliminate HCl gas in a preliminary step at 300-400 °C
before more extensive cracking reactions take place. The
dry HCl gas would then be available for further use.

The residual content of organic chlorine is of great
importance for the value of the pyrolysis oils. For processing in existing petrochemical plants the chlorine
content should not exceed 10 ppm. Oils that are obtained
from the pyrolysis of plastic mixtures contain between 50
and 200 ppm of organically bound chlorine. Fortunately,
no chlorinated dibenzodioxines were found among these
chloro-organic compounds.

A further dehalogenation of the pyrolysis oil was
achieved by dosing sodium vapour into the flow of pyrolysis gas.

A semi-industrial plant using the Hamburg fluidized
bed pyrolysis process was built in Ebenhausen near Munich
and operated by the Asea Brown Boveri company (ABB) with
a throughout of 5000 t/a.

The process has been shown to be successful for the
pyrolysis of polyolefins.

After experiments on the pyrolysis of plastic wastes
had shown how individual pieces that were only slightly
smaller than the reactor's inner diameter could be pyrolyzed in the fluidized bed, a larger pilot plant with a
90 cm square fluid bed was used to decompose whole used
tires (Figure 2). The tires were introduced to the reactor through a lock with pneumatic gates. The steel parts
remaining in the reactor, after pyrolysis was complete,
were forked out of the fluidized bed by means of a grate

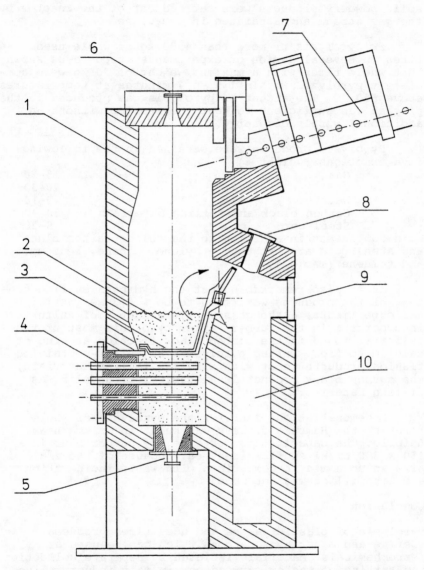

Figure 2. Flow Diagram of the Prototype Reactor for Whole-Tire
Pyrolysis. (1) steel wall with fireproof bricking; (2) fluidized
bed; (3) tiltable grate; (4) radiation fire tubes; (5) nozzles to
remove sand and metal; (6,8,9) flange for observation and re-
pairs; (7) gastight lock; (10) shaft for steel cord.

at programmable intervals and deposited in a silo. The
solid powdery products were carried out of the reactor by
the gas stream and separated in a cyclone.

By pyrolysis of more than 4500 kg of whole used
tires in several series of experiments it could be shown
that whole tires with a maximum weight of 20 kg were completely pyrolyzed within 1,5 to 4 minutes at temperatures
above 650 °C. The amount of pyrolysis gas produced in the
process was sufficient to heat the reactor without an
additional supply of energy.

Pyrolysis products were obtained in the following
five fractions (weight %):

Gas	15-20
Oil	20-30
Water	5-10
Carbon black and filling materials	30-40
Steel cord	5-20

Main components in addition to the solids carbon black
and steel cord are methane, ethylene, ethane, benzene,
and toluene (Table 4).

It is also remarkable that the hydrogen sulfide content of the products was found to be < 0,3 weight % in
all experiments, although some 2,0 weight % of sulfur are
incorporated in the rubber. After pyrolysis most of the
sulfur is found in the carbon black fraction as zinc or
calcium sulfide. During pyrolysis the sulfur-containing
fission products react with alkaline filling materials.
The carbon black product is suitable for recycling as
filling material.

A demonstration plant for the pyrolysis of whole
tires by the Hamburg process was built in Grimma near
Leipzig. The capacity was 100 tire pieces per hour in a
150 x 300 cm sized fluidized bed. Because of the small
size an economic in comparison with waste incineration
and landfilling was not made.

Conclusion

Pyrolysis of plastic wastes and used tires produces
olefins and other hydrocarbons which can be used as
petrochemicals. Materials isolated by separate collection
and sorting of wastes carried out in many industrialized
nations are suitable for recycling through pyrolysis.
Because olefins and oil are generates as valuable
substances, the profitability of the process depends
largely on the price of petroleum. The combination of a
refinery together with a pyrolysis plant would optimize
the use for the products. To minimize the costs, plastic
wastes have to be collected in larger quantities so that
bigger plants can be built. Some of these processes are

likely to be established in the market in the near future.

Literature Cited

1. Albright, L.F.; Crynes, B.L.; Corcoran, W.H. *Pyrolysis Theory and Industrial Practice*; Academic Press: New York, 1983
2. Hawkins, W.L., *Polymer Degradation and Stabilization*; Springer Press: Berlin, 1984
3. Ferrero, G.L.; Maniatis, K.; Buekens, A.; Bridgewater, A.V., *Pyrolysis and Gasification*; Elsevier A. Science: London, 1989
4. Bracker, G.P., *Pyrolysis*, in Ullmanns Enzyklopädie der Technischen Chemie, 4th ed., vol. 6, Verlag Chemie: Weinheim, p. 553
5. Thomé-Kozmiensky, K.J., *Pyrolyse von Abfällen*; EF-Press: Berlin 1985
6. Sinn, H.; Kaminsky, W.; Janning, J. Angew.Chem. 1976, *88*, 737; *Angew.Chem.Int.Ed.Engl.* 1976, *15*, 660
7. Piskorz, J.; Radlein, D.; Scott, D.S., *J.Analy.Appl. Pyrol.* 1986, *9*, 121
8. Schulten, H.-R.; Lattimer, R.P.; Boon, J.J., Eds.; Pyrolysis, *J.Anal.Appl.Pyrolysis* 1991, *19*
9. Sinn, H., *Chem.-Ing.-Tech.* 1974, *46*, 579
10. Kaminsky, W.; Sinn, H.; Janning, J., *Chem.-Ing.-Tech.* 1979, *51*, 419
11. Ricci, L.J., *Chem.Eng.*, Vol. 83, No. 16, 52, 1976
12. Wenning, H.P., *Chem.-Ing.-Tech.* 1989, *61*, 277
13. Nowak, F., *Energie* 1978, *7*, 186
14. Kaminsky, W., *Ressour. Recovery Conserv.* 1980, *5*, 205
15. Kaminsky, W., *J.Anal.Appl. Pyrolysis* 1985, *8*, 439
16. Matsumoto, K.; Kurizu, S.; Oyamoto, T.. In *Conversion of Refuse to Energy*; Montreux 3-5 Nov 1975, Conf. Papers IEEE Catal. no. 75, Eiger AG, Zurich, 538
17. Sueyoshi, H.; Kitaoka, Y., *Hydrocarbon Process* 1972, 161
18. Schulmann, B.; White, P.A., *Pyrolysis of Scrap Tires Using the Tosco II-Process*, ACS Sympos. Ser. 1978, 76
19. Kaminsky, W.; Menzel, J.; Sinn, H., *Resour.Recovery Conserv.* 1976, *1*, 91
20. Kaminsky, W.; Sinn, H., *Hydrocarbon Process* 1980, *59*, 187
21. Kaminsky, W.; Kummer, A.B., *J.Anal.Appl.Pyrolysis* 1989, *16*, 27
22. Kaminsky. W., *Makromol.Chem.Macromol.Symp.*1991, *48/49*, 381
23. Buekens, A.; de Wolf, F.; Schoeters, J. In *Pyrolysis and Gasification*; Ferrero, G.L., Maniatis, K., Buekens, A., Bridgwater, A.V.; Elsevier Science, London, 1989; 580
24. Kaminsky, W.; Franck, J., *J.Anal.Appl.Pyrolysis* 1991 *19*, 311

RECEIVED August 26, 1992

STABILIZERS, ADDITIVES, AND CHARACTERIZATION

Chapter 7

Stabilization of Recycled Plastics

Peter P. Klemchuk and Tom Thompson

Additives Division, Ciba-Geigy Corporation, Ardsley, NY 10502

As concerns about the disposal of increasing quantities of municipal solid waste (MSW) have grown, many approaches to solid waste disposal are being explored: source reduction, recycling, composting, incineration, degradable plastics and sanitary landfills. Recycling is experiencing a "grass roots" upsurge of interest and, although much remains to be accomplished, especially to increase effectiveness of collection and reduce costs of recycled materials, recycling is likely to be a major means of resource recovery and MSW reduction. The recycling of paper, metals, and glass has been an ongoing activity for some time. The recycling of plastics is relatively new but showing signs of becoming an important means for reducing the amounts of plastics in MSW. Plastics are unique among recoverable materials in that they require stabilizers for processing and fabrication and to withstand end-use conditions. This paper presents some information regarding the needs of plastics for stabilization and proposes some strategies for dealing with the stabilization needs of recovered plastics. Findings from a limited study of the stabilization of recycled high density polyethylene are presented. All-in-all, published results from earlier studies and the results from this study indicate that recycled plastics can be stabilized adequately to meet performance requirements of new applications. However, it should be pointed out that much more must be learned about the stabilization of recovered plastics before routine practices for the purpose can be established.

Solid waste disposal is an important issue in the United States because of the increasing volume of MSW and the decreasing landfill capacity for disposing of it. Source reduction, recycling, composting, and incineration are receiving attention as alternatives to landfills. Heightened awareness of the solid waste problem has led to much activity in the political arena and to multiple approaches in research to solve the problem. The main political activities have shifted from the federal government in Washington to

0097–6156/92/0513–0074$06.00/0
© 1992 American Chemical Society

individual states and cities. The problem is receiving much "grass roots" attention. This has had the effect of bringing about changes in the way municipal trash is handled at the local level. Many recycling programs have been started which recycle paper, metal, plastics and other materials. This recycling renaissance is expected to gain momentum and become a major feature of solid waste disposal programs. Recycling of plastics is expected to be an important part of that activity.

Among recyclable materials - paper, metals, glass, plastics - plastics are unique in being prone to thermo-oxidative and photo-oxidative degradation. They require small amounts of stabilizing additives to retard the loss of properties and to find practical uses. In addition, they are dissimilar chemically and morphologically, as are metals, and are most useful and valuable when homogeneous and not mixed. Sorting plastics economically, developing strategies for reusing recovered materials, especially regarding stabilization needs and applications, and finding markets for recovered plastics are three big needs for a successful plastics recycling program.

This paper reviews the status of plastics recovery from MSW, the future prospects for plastics recovery, some problems associated with recycling plastics, some perspectives on the need of plastics for stabilization, and some proposed strategies for the stabilization of recovered plastics. Also included are some limited results from our own work and that of others on the degradation and stabilization of recycled high density polyethylene (HDPE) and polyethylene terephthalate (PET).

The need for moving quickly on solid waste problems has resulted in recycling running ahead of investigations of the stabilization of recovered plastics. However, information from stabilization practices with virgin plastics can be used in the meantime, until studies with recovered materials are available, to develop meaningful strategies for the stabilization of recovered plastics. Not until studies with recovered materials are completed will it be possible to be comfortable with current practices for stabilizing recovered plastics. Even then, there may be instances when unexpected situations may be experienced due to unusual materials being recycled, especially materials of unknown history which may have included exposure to severe conditions.

Plastics in Municipal Solid Waste

Tables 1 and 2 contain information on the breakdown of MSW in the United States in 1988 according to material type by weight and volume and by product type, respectively. Containers, packaging and nondurable goods were almost 60% of MSW in 1988 (Table 2). Of the nearly 180 million tons of MSW in 1988 (from 88 million tons in 1960), 73% was disposed by landfill, 14% by incineration, and 13% by recovery. Paper and paperboard have been the largest contributors to MSW from 1960 to the present. Yard wastes, much of which could be degraded by composting, are in second place. Plastics, in fourth place after metals, were 8% of the MSW weight and almost 20% of the MSW volume.

As shown in Table 3, most of the 13.1% by weight of MSW recovered in 1988 by recycling and composting were paper, metals and glass. Only 1.1% of the plastic component was recycled and that was mainly polyethylene terephthalate from soft drink bottles. A study by Franklin Associates Ltd.(2)predicts plastics will increase to 9.3% by weight of MSW by 1995, from 8.0% in 1988. The same study predicts by 1995 significant reductions in quantities of MSW discarded in landfills, significant increases in amounts incinerated and also increases in amounts recovered. The portion of plastics which is recovered from MSW is expected to be between 2.7 and 6.7% in 1995, up from 1.1% in 1988 (Table 3).(2)   The amounts being recovered presently have increased significantly over 1988.

TABLE 1
MUNICIPAL SOLID WASTE IN THE UNITED STATES BY MATERIAL
TYPE, 1988; WEIGHT AND VOLUME

Material	WEIGHT		VOLUME	
	Million Tons	%	Million Cu. Yds.	%
Paper	71.8	40.0	136.2	34.1
Yard Wastes	31.6	17.6	41.3	10.3
Metals	15.3	8.5	48.3	12.1
Plastics	14.4	8.0	79.7	19.9
Food Wastes	13.2	7.4	13.2	3.3
Glass	12.5	7.0	7.9	2.0
Other*	20.8	11.6	73.4	18.4
Total:	179.6		400.0	

* rubber, leather, textiles, wood, misc., inorganic wastes, other
SOURCE: Adapted from ref. 1.

TABLE 2
MUNICIPAL SOLID WASTE IN THE UNITED STATES
BY PRODUCT TYPE 1988

Product	Million Tons	%
Containers, Packaging	56.8	31.6
Nondurable Goods	50.4	28.1
Yard Wastes	31.6	17.6
Durable Goods	24.9	13.9
Food Wastes	13.2	7.4
Other	2.7	1.5
Total:	179.6	

SOURCE: Adapted from ref. 1.

TABLE 3
RECOVERY OF MSW IN THE UNITED STATES, 1988 AND 1995

Material	Generated, Million Tons		% Recovered		
	1988	1995	1988	1995Lo	1995Hi
Paper	71.8	85.5	25.6	30.8	38.1
Yard Wastes	31.6	33.0	1.6	20.0	33.3
Metals	15.3	16.2	14.6	21.0	29.0
Plastics	14.4	18.6	1.1	2.7	6.7
Food Wastes	13.2	13.2	Neg.	0.0	7.6
Glass	12.5	11.1	12.0	18.9	27.9
Other	20.8	17.2	3.8	8.4	12.7
Total:	179.6	199.8	13.1	20.0	27.7

SOURCE: Adapted from ref. 1.

The merits of recycling materials have been recognized and there is considerable ac
tivity in the United States in support of recycling. At the time of this writing, 32 state
have passed laws to encourage and increase recycling. More than 1500 communities i
35 states have established MSW collection programs to support recycling.(*3*) Othe
states and communities are expected to do the same.

Recycling of Plastics in Municipal Solid Waste

An estimated 61% of the top 15 plastic resins produced in the United States in 198
was disposed in MSW (Table 4). The largest portion of the total, 59.7%, came fron
residential sources, 25% from commercial sources and 15.3% from institutions. Onl
about 1.1% of the plastics in MSW in 1988 was recycled and that was mainly recover
from recycling 28% of polyester soft drink bottles. Significant increases in the recy
cling of plastics are expected by 1995 (Table 3). To facilitate recycling, the Society o
the Plastics Industry (SPI) has developed a voluntary code to identify the plastic use
to make an article. The adoption and use of these codes by plastic fabricators shoul
increase the amounts of sorted recycled plastics and increase the value of these materi
als. Recycling of polyester soft drink bottles has been the biggest success so far in th
reuse of plastic materials from MSW. Major reasons for the success are the availabilit
of a source of homogeneous plastic and of marketing outlets for the recovered material
In essence, recycling transforms polyester bottles into fiber for carpeting, outerwear in
sulation, etc.

TABLE 4
SUMMARY OF 1988 RESIN PRODUCTION & DISPOSAL

| | Production | Quantity Disposed | | | |
| | | Non-MSW | | MSW | |
	MM lb	MM lb	%	MM lb	%
Acrylic	686.0	663.0	96.7	22.7	3.3
Acrylonitrile-butadiene-styrene	1,093.3	383.1	35.0	710.2	65.0
High-density polyethylene	6,528.8	975.3	14.9	5,553.5	85.1
Low-density polyethylene	7,690.8	577.2	7.5	7,113.6	92.5
Nylon	461.6	329.2	71.3	132.4	28.7
Phenolic	2,975.1	2,869.0	96.4	106.1	3.6
Polyethylene terephthalate & Polybutylene terephthalate	1,475.5	176.2	11.9	1,299.3	88.1
Polypropylene	5,274.0	1,016.9	19.3	4,257.1	80.7
Polystyrene	4,767.9	529.7	11.1	4,238.2	88.9
Polyurethane	2,794.8	1,510.7	54.1	1,284.1	45.9
Polyvinyl chloride	7,566.0	5,799.4	76.7	1,766.6	23.3
Unsaturated polyester	1,319.3	1,183.0	89.7	136.3	10.3
Urea & melamine	1,459.2	1,346.7	92.3	112.5	7.7
Total:	44,092.3	17,359.7	39.4	26,732.6	60.6

SOURCE: Adapted from ref. 2.

The Coca-Cola Company (Hoechst Celanese) and PepsiCo, Inc. (Goodyear) have an
nounced plans to recycle the polyester used in their bottles. The former will use a

process which recovers dimethyl terephthalate, the major starting material for polyethylene terephthalate, by methanolysis of the soft drink bottles. The recovered, purified monomer will be used in polymerizations to make virgin polymer for bottles. The later will use a process which recovers bis(hydroxy-ethyl) terephthalate, also a starting material for PET, by glycolysis of the polymer. Coca-Cola/Hoechst Celanese have received approval from the FDA to use the polymer from their recovery process for food packaging.

Major barriers exist to the large scale recycling of plastics. Among them are the infancy of the activity so that the infrastructure and procedures for efficient, large-scale recycling are not yet in place - getting waste plastic articles from sources to plastics recyclers is far from smooth and well organized. News reports of problems encountered in implementing state-wide recycling in Connecticut on January 1, 1991 emphasize the startup difficulties of new recycling programs - most communities had not been able to provide in time the necessary facilities for collection, sorting, and storage of materials to be recovered from MSW.(4) End-use applications for recycled plastics are far from optimum. With the exception of polyester bottles, sufficient means haven't existed until now for collecting and recycling sorted streams of plastics - i.e. all HDPE, all PP, all LDPE, etc. Such homogeneous streams would provide a higher value material for recycling into end products than the so-called commingled streams which are used for plastic "lumber" for picnic tables, park benches, marine facilities, etc. As recycling increases, as the necessary facilities and procedures are in place and working smoothly, these barriers will disappear. But, meanwhile, plastics recycling is far from a humming business activity.

Economics and legislation will play important roles in determining the success of recycling. Legislation at the state and community levels has already led to the establishment of many municipal recycling programs. Legislation at the federal and state levels may be foreseen to mandate the use of recycled materials with virgin materials in a variety of items purchased by federal and state governments. However, recycling costs will continue to be a major factor. In a recent example, the federal government found the cost of tax forms printed on recycled paper would be as much as 42% more costly than forms printed on virgin paper, in spite of the cost of recycled newspapers having dropped by 80% to $1.30 a ton.[3]

Prospects for The Stabilization of Recycled Plastics

The major polymers to be considered for recycling are PET, HDPE, polypropylene (PP), polystyrene (PS) and polyvinyl chloride (PVC). The data in Table 4 indicate more than 80% of all but PVC were disposed in MSW in 1988; 23% of PVC went into MSW. The data in Table 5 show only small quantities of these plastics sold in 1989 were recovered but that the potential for recycling into non-food applications is much greater and is very likely to increase with time as recycling becomes better established.

Applications identified for recovered plastics disclose a mixture of low stress and high stress (weathering) uses.(5) The low stress applications will require minimal stabilization, usually low concentrations of processing stabilizers, often trivalent phosphorus compounds and phenolic antioxidants, to protect the plastics during fabrication and end-use. Weathering applications require the use of light stabilizers in addition to processing stabilizers and phenolic antioxidants. The stabilization

TABLE 5
SALES, ACTUAL RECOVERY, AND POTENTIAL RECOVERY OF
MAJOR POLYMERS IN MSW, 1989, MILLION LBS.

Polymer	Sales	Recovered (%)		Potential Non-Food Recovery (%)	
PET	1905	190	(10)	529	(28)
PET*	680	190	(28)	529	(78)
HDPE	8100	145	(1.8)	440	(5)
PP	7200	62	(0.9)	770	(11)
PVC	5000	5	(0.1)	490	(10)
PS	5200	20	(0.4)	480	(9)

*soft drink bottles
SOURCE: Adapted from ref. 5.

requirements of virgin plastics named in Table 5 are varied. PVC is usually stabilized with metal salts for thermal stability - those stabilizers are not used for the other polymers. PET is usually stabilized with small amounts of trivalent phosphorus compounds for stabilization during processing and to minimize discoloration caused by catalyst residues from polymerization. Virgin polyolefins and virgin PS are the biggest consumers of antioxidants used in polymers; polyolefins are the biggest consumers of light stabilizers used in polymers.(6)

The stabilization requirements of recycled plastics in Table 5 are at least as varied as those of the virgin materials. Limited information is available on the stabilization requirements of recovered plastics. Until more is known about those needs and until users of recovered plastics have more information to guide them, it seems reasonable to suggest using stabilizers in amounts which would be used in virgin materials for the same applications. This suggestion deserves consideration especially since recycled materials come from a variety of sources with a variety of histories. The one exception to this suggestion may be the use of thermal stabilizers for PVC since those stabilizers are already used at levels usually greater than 1%, and lesser amounts may be sufficient for reuse of recovered PVC. Table 6 lists the concentrations and types of stabilizers used for the stabi;lization of virgin hydrocarbon polymers. The information can be used as a guide in selecting stabilizers and concentrations for all polyolefin and PS applications and for weathering applications of PVC.

Published Information on The Stabilization of Recycled Plastics

S. Dietz(7) in a paper, "The Use and Market Economics of Phosphite Stabilizers in Post Consumer Recycle" presented data which showed phosphite stabilizers were effective in providing color stability and melt stability in virgin PET during processing at 280°C. Phosphite stabilizers were effective also in providing color stability in post consumer recycled PET under the same processing conditions. Higher concentrations, from 0.2% to 0.5%, were used in the recycled PET whereas lower concentrations, 0.1 to 0.2%, were used in virgin polymer for color stability.

The reuse of PET recovered from soft drink bottles in a number of new, non-food applications is already a success story. The new uses can be achieved with moderate amounts of stabilizers, partly because the quantities required for stabilization of virgin

TABLE 6
STABILIZERS AND CONCENTRATIONS USED AT VARIOUS
STAGES IN HYDROCARBON POLYMERS

Stage	Stabilizer	Usual Concentration
• Drying	antioxidant	<250 ppm
• Storage	antioxidant	<250 ppm
• Compounding	antioxidant +	500-1000 ppm
	phosphite	500-1000 ppm
• Fabrication	same as Compounding	
End-use:		
• Low stress	same as Compounding	
• Thermal stress	antioxidant +	1000-5000 ppm
	thiosynergist +	1000-5000 ppm
	phosphite	500-1000 ppm
• Weathering	antioxidant +	0-1000 ppm
	phosphite +	500-1000 ppm
	hindered amine +	0-10000 ppm
	ultraviolet absorber	0-10000 ppm

resin are low, and partly because the new applications are not highly stressing, as for example, weathering would be.

Stabilization studies reported with recycled PP and HDPE indicate these polymers also respond to stabilizers so that they can be used for new applications, even stressful ones, as long as adequate stabilization is provided.[8] The addition of 0.05% of a phosphite processing stabilizer at every stage in the multiple extrusion of recycled PP, i.e., at the beginning or after one, two, three or four extrusions, was effective in providing melt stabilization. These results mean that, no matter how much processing history the unstabilized recycled PP had experienced, the addition of a phosphite processing stabilizer provided stabilization during subsequent melt processing.

In the same study,[8] PP samples, which had been aged in an oven at 135°C for 35 days without failure, and then remolded, were found to survive for 17 more days at 135°C before failing. However, when 0.15% of a 2:1 blend of phosphite processing stabilizer and hindered phenolic antioxidant was added to the reprocessed and remolded polymer the additional lifetime was 100 days at 135°C. This result is also encouraging regarding prospects for improving the lifetimes of recycled plastics by the addition of stabilizing additives.

HDPE bottle crates, samples of which had lasted 8000 hours to failure by cracking in a xenon arc weatherometer when first made, were found after five years' service in a weathering environment to have dropped in stability to 1000 hours to failure by cracking. However, the addition of 0.1% of a hindered amine light stabilizer to reformulated and remolded crates increased the time to failure by cracking to original values - 8000 hours![8]

Investigation of The Stabilization of Recycled High Density Polyethylene

Experimental.

Materials. The processing stabilizer used in this study was tris(2,4-di-*tert*-butyl phenyl)phosphite; the hindered phenolic antioxidant was tetrakis[methylene(3,5-di-*tert* butyl-4-hydroxyhydrocinnamate)]methane; the hindered amine light stabilizer wa 2,2,6,6-tetramethyl-4-piperidyl sebacate. Commercial qualities of these additives wer used.

The injection molding grade of HDPE was Alathon 5046. The virgin HDPE bottl resin was obtained from a major commercial producer of HDPE bottle resins. Th post-consumer recovered HDPE was obtained from the Rutgers University Recyclin Research Center. The pre-photo-oxidized HDPE polymer was prepared by exposing 8 x 8" x 0.005" sheets of compression molded polymer in an exposure device to radiatio from Atlas UVA-340 fluorescent sunlamps.

Procedures. Polymer samples were ground to fine powders in a Fritsch powde mill. Formulations were prepared by adding stabilizing additives dissolved in methyl ene chloride to powdered polymers and blends of polymers which contained powdere calcium carbonate. The solvent was removed in a hood and the formulations wer dried to constant weight. The thoroughly blended formulations were compressio molded at 190°C in a picture-frame mold of the requisite dimensions. Molding time were 2 min. contact at low pressure, 1 min. contact at high pressure, followed by cool ing for 2 min. at high pressure. The resultant compression molded sheets were cut int pieces which were less than 0.25 in. and were compression molded again under th same processing conditions into sheets which were 0.005" thick.

Photo-oxidation was carried out in a homemade exposure device in which forty 48 UVA-340 fluorescent sunlamps are mounted vertically in the form of a cylinder with diameter of 34". Samples were mounted vertically 2" away from the bulbs on alumi num shelves which were rotated at 1.25 rpm. Samples for tensile measurements wer precut to 3" x 0.25"x 0.005". Samples for carbonyl absorbance measurements by infra red were 0.005" thick and were mounted on standard film holders for infrared meas urements. Tensile measurements were carried out on an Instron instrument. Carbony absorbance values at 1710 cm^{-1} were calculated from absorption spectra by subtractin absorbance at the reference wavelength of 1880 cm^{-1}.

Thermal oxidation was carried out in a Blue M Power-O-Matic 60 circulating ai oven at 120°C. Samples prepared for oven aging to failure, 2" x 1"x 0.005", were ex amined weekly. Samples for infrared measurements were mounted in the same type o film holders as the samples for photo-oxidation.

Results and Discussion. At the beginning of this work on the stabilization of recycle HDPE several 8" x 8" x 0.005" sheets of an injection-molding grade of HDPE wer photo-degraded by exposure in a device, 340 FL, containing fluorescent lamps wit emission centered at 340 nm and no humidity control or water spray.

Property changes of the HDPE samples as a function of exposure time are summa rized in Table 7. The carbonyl absorbance of 0.20 after 873 hours exposure, the sever reductions in tensile elongation, energy-to-break and molecular weight, all indicate ex tensive degradation of the HDPE had taken place. This photo-degraded HDPE was considered to be representative of polymer which had undergone severe photo-degradation but still retained enough physical integrity to be suitable for recovery. It was considered a "worst-case" recycled resin. This resin was evaluated, as is, and in combination with the virgin injection molding resin from which it was derived. Unstabilized and stabilized formulations were evaluated for thermo-oxidative stability in oven aging at 120°C and for photo-stability in a 340 FL exposure device.

TABLE 7
PHOTO-DEGRADATION OF UNSTABILIZED INJ. MOLD. HDPE
PREPARATION OF WORST-CASE "RECYCLED" RESIN

	HOURS 340 FL EXPOSURE			
PROPERTY	0	355	638	873
CARBONYL ABS.	0.012	0.044	0.096	0.202
ELONGATION (%)	1881	1945	1691	438
STD. DEV. (%)	12	8	35	48
ENERGY-TO-BREAK (KG-MM)	1031	1027	940	79
STD. DEV. (%)	22	14	2	25
MOL. WEIGHT. (Mn)	21,500	20,700	17,500	15,200
MOL. WEIGHT. (Mw)	117,000	113,000	87,800	63,600
DISPERSION	5.4	5.5	5.0	4.2

A sample of post-consumer HDPE recovered from water bottles and milk bottles was obtained from the Rutgers University Recycling Research Center. It was evaluated alone and in with virgin resins in unstabilized and stabilized formulations for thermal and photo-stability.

In total, four samples of HDPE were included in the study (Table 8). One was the virgin injection molding grade which had been used to prepare the second sample, the "worst-case" recycled polymer by photo-degradation. The third sample was the post consumer resin recovered from unpigmented bottles. The fourth was a sample of a virgin blow molding grade of HDPE for bottles.

Particularly striking among the four resins is that the percent elongations of the virgin and recycled bottle resins were much lower than that of the injection molding resin. The elongation and energy-to-break of the recycled bottle resin were particularly low and reproducibility was poor. That may have been due to trace amounts of indiscernible impurities or it may have been due to low stabilization levels of water and milk bottle resins which do not permit the resins to survive the compounding, bottle blowing, use and recycling experiences with minimal change. That result suggests the use of higher stabilizer levels in the original bottle resins should be investigated to determine if they would provide recycled resin with good properties and good reproducibility of properties. If such is the case, these recycled resins might not need additional stabilizers to provide adequate performance for non-demanding reuse applications.

Although the prephoto-oxidized resin in Table 9 had suffered a significant reductio
in properties as a result of the exposure to uv light, the photo-stabilized compositions
Table 9 show that blends of the virgin and photo-oxidized resins exhibited high leve
of elongation, comparable to the virgin resin, and low levels of standard deviatio
These results mean that even when 10-25% of a highly photo-degraded polymer is

TABLE 8
PROPERTIES OF HDPE RESINS IN STUDY

PROPERTY	HDPE RESINS			
	INJ.MOLD. VIRGIN	INJ.MOLD. PHOTOOXID.	BOTTLE VIRGIN	BOTTLE RECYCLED*
BREAKING STRENGTH (PSI)	6040	1080	2930	2280
STD. DEV. (%)	45	23	31	11
ELONGATION (%)	1754	512	265	60
STD. DEV. (%)	3	22	104	108
ENERGY-TO-BREAK (KG-MM)	940	79	206	50
STD. DEV. (%)	12	25	85	90
MELT FLOW**	1.01	1.38	0.85	0.77
AO ANALYSIS	350 ppm	-	100 ppm	75 ppm
MOL. WEIGHT. (Mn)	21500	15200	16000	17000
MOL. WEIGHT. (Mw)	117000	63600	130000	130000
DISPERSION	5.4	4.2	8.1	7.6

*Rutgers University Recycling Research Center
**190 C, 2.16 kg., 0.25 in.

TABLE 9
IMPACT OF *PREPHOTOOXIDIZED* HDPE ON PHOTOSTABILITY OF VIRGIN HDPE
340 FL EXPOSURE; 5 MIL FILMS; GEOMETRIC MEANS OF 5-7 REPLICATES

% POLYMER		%ELONGATION @ HOURS EXPOSURE*						
VIRGIN**	OXID.	0	233	496	687	973	2509	4093
100	0	1664	1350	1528	1531	1479	1455	1240
90	10	1457	1255	1535	974	1436	799	365
75	25	1582	1289	1505	1426	852	164	102
50	50	1234	1028	800	1121	412	49	55
0	100	461	401	139	111	26	20	14

% POLYMER		%STANDARD DEV. @ HOURS EXPOSURE						
VIRGIN**	OXID.	0	233	496	687	973	2509	4093
100	0	12	18	10	17	10	11	38
90	10	23	23	6	51	7	36	177
75	25	9	7	5	12	48	94	132
50	50	21	31	20	15	127	32	47
0	100	83	16	204	86	31	26	17

*Stabilizer Composition: 750 ppm Ca stearate, 250 ppm proc. stab., 1000 ppm HALS

combined with virgin polymer the resultant composition can have acceptable properties as long as appropriate stabilization is used.

The same conclusion applies to the data in Table 10 for a virgin injection molding grade of HDPE and bottle resin recovered at a post consumer recycling facility. The result is the more remarkable since the recycled bottle resin has relatively very low elongation and poor reproducibility of elongation. These results suggest it may not be necessary to limit the use of recycled HDPE resins to formulations with the same grade of resin. Satisfactory performance appears likely from compositions of mixed resin grades, as long as the proportion of recycled polymer in the composition does not exceed much more than 25%.

TABLE 10
IMPACT OF *RECYCLED* HDPE ON PHOTO-STABILITY OF VIRGIN HDPE
340 FL EXPOSURE; 5 MIL FILMS; GEOMETRIC MEANS OF 5-7 REPLICATES

% POLYMER		%ELONGATION @ HOURS EXPOSURE*						
VIRGIN**	RECY.***	0	233	496	687	973	2509	4093
100	0	1664	1350	1528	1531	1479	1455	1240
75	25	1192	1302	1432	1270	1486	1161	764
0	100	86	101	102	104	90	31	30

% POLYMER		%STANDARD DEVIATION @ HOURS EXPOSURE						
VIRGIN**	RECY.***	0	233	496	687	973	2509	4093
100	0	12	18	16	17	10	11	38
75	25	40	11	10	23	7	25	54
0	100	86	125	281	99	182	82	75

*Stabilizer Composition: 750 ppm Ca stearate, 250 ppm proc. stab., 1000 ppm HALS
**Injection Molding Grade
***Rutgers University Recycling Research Center

The plots of elongation versus carbonyl absorbance in Figure 1 for a set of 5 mil HDPE formulations, which had been exposed in a 340 FL exposure device, indicated that a carbonyl absorbance of 0.2 would be a conservative index for failure of photo-degraded samples. Since the carbonyl absorbances of photo-stabilized formulations showed virtually no change over more than 6000 hours' exposure in the 340 FL exposure device and since only a limited number of samples was available for tensile property testing, a carbonyl absorbance of 0.2 was selected as the criterion of failure for photo-oxidizing samples. Exposure times to generate 0.2 carbonyl absorbance of the 5 mil films exposed in the 340 FL exposure device are summarized in Table 11. Whereas unstabilized films containing various amounts of recycled or photo-degraded HDPE failed in a few hundred hours, those stabilized with 250 ppm of a processing stabilizer plus 1000 ppm of a hindered amine light stabilizer required more than six thousand hours to reach a carbonyl absorbance of 0.2. These results suggest that recycled hydrocarbon plastics will respond well to stabilization and be suitable for many new demanding applications.

Results in earlier work(9) for samples of HDPE which had been aged in an oven at 100°C showed that the unstabilized 6 mil films in that study had become brittle when increases in carbonyl absorbances of 0.4 had been reached. The stabilized films had become brittle with carbonyl absorbances of 0.15. Therefore, for the present study, the failure index for oven aging at 100°C was an increase in carbonyl index of 0.15 for

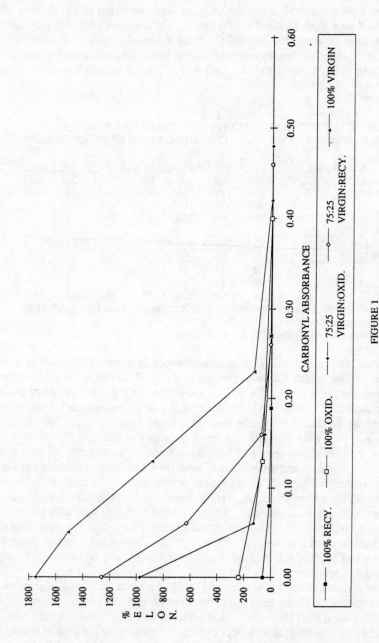

CARBONYL ABSORBANCE

100% RECY.   100% OXID.   75:25 VIRGIN:OXID.   75:25 VIRGIN:RECY.   100% VIRGIN

FIGURE 1
% ELONGATION VS CARBONYL ABSORBANCE OF
THERMALLY STABILIZED 5MIL HDPE FILMS
340 FL EXPOSURE; PHOTO-DEGRADATION

TABLE 11
PHOTO-STABILITY (340 FL) OF UNSTABILIZED AND STABILIZED 125 MICRON HDPE FILMS
COMPARISON OF FOUR HDPE RESINS AND 75:25 BLENDS

POLYMER COMPOSITION (%)				HRS. TO 0.2 CARB. ABS.	
INJ. MOLD. VIRGIN	INJ. MOLD. PHOTO-DEG.	BLOW MOLD. VIRGIN	BOTTLE RECYCLED*	UNSTAB.	PHOTO-STAB.**
100				690	>6400
	100			260	>6400
		100		360	>4900
			100	480	>6400
75	25			460	>6400
		75	25	250	>4900

Post-consumer recycled water and milk bottles from Rutgers Univ. Recycling Research Center.
*750 ppm Ca stearate, 250 ppm proc.stab., 1000 ppm HALS

thermally stabilized films and 0.4 for unstabilized films. Plots of carbonyl absorbance versus exposure time were interpolated to provide the times to failure which are summarized in Table 12.

The results in Table 12 show the significant increases in thermal stability which had been provided by the thermal stabilizer formulation. As with the results in Table 11 for photo-stabilization, the results in Table 12 for thermal stabilization are very encouraging regarding the prospects for stabilizing recycled HDPE for new uses.

TABLE 12
THERMAL-STABILITY (Oven, 120°C) OF UNSTABILIZED AND STABILIZED HDPE FILMS
COMPARISON OF FOUR HDPE RESINS AND 75:25 BLENDS; 125 MICRON FILMS

POLYMER COMPOSITION (%)				HRS. TO FAILURE**	
INJ. MOLD. VIRGIN	INJ. MOLD. PHOTO-DEG.	BLOW MOLD. VIRGIN	BOTTLE RECYCLED*	UNSTAB.	THERM. STAB.***
100				720	1850
	100			100	930
		100		770	2060
			100	320	2090
75	25				1440
		75	25		1850
75			25		2020

* Post-consumer recycled water and milk bottles from Rutgers Univ. Recycling Research Center.
**Failure indices: Carb. Abs. +0.4 for unstabilized and +0.15 for stabilized formulations.
***750 ppm Ca stearate, 500 ppm phenolic AO, 500 ppm proc. stab.

Conclusions

- Recycling of polyethylene terephthalate from soft drink bottles and high density polyethylene from water and milk bottles are currently the most advanced post consumer recovery operations for plastics. The value of these two post consumer recovered plastics is relatively high because of ease of sorting and high quality. PET and HDPE have at present the best prospects for commercial success as recycled plastics.

- The performances of 75:25 virgin:recycled formulations with HDPE were found to be satisfactory for most uses, even when pre-photo-oxidized HDPE was used in the formulations.

- Results of this study show prospects are very good for the stabilization of post consumer HDPE to achieve high levels of thermal stability and photo-stability in formulations with virgin resins.

- The reduction in properties of HDPE recovered from water and milk bottles, as compared with virgin bottle resins, suggests the use of higher levels of stabilizers should be studied in the virgin bottle resins to determine if better properties of recovered HDPE will result.

- Published results, and the results of this study, show post-consumer PET and HDPE respond well to stabilizers. Information is not yet available for other recovered polymers. Stabilization studies should be carried out with all recovered plastics to establish the types and levels of stabilizers needed for new uses. In the meantime, the stabilization practices for virgin polymers are good guides to apply to recovered plastics for comparable uses.

Literature Cited

1. U.S. Environmental Protection Agency, *Characterization of Municipal Solid Waste: 1990 Update*, EPA/530-SW-90-042, June 1990.
2. Franklin Associates Ltd., *Characterization of Plastic Products in Municipal Solid Waste*, January, 1990, Council for Solid Waste Solutions.
3. Thayer, A. M., "Degradable Plastics Generate Controversy in Solid Waste Issues", *Chemical & Engineering News*, June 25, 1990, p.7.
4. *The New York Times*, January 22, 1991, p. B1.
5. Bennett, R.A., Technical Report #41, "New Product Applications and Evaluations & Continued Expansion Of A National Database On The Plastics Recycling Industry", The Center for Plastics Recycling Research, Rutgers, The State University of New Jersey, 1991.
6. Chemical Economics Handbook-SRI International, *Plastics Additives*, 1985.
7. Dietz, S., "The Use and Market Economics of Phosphite Stabilizers in Post Consumer Recycle" presented at RECYCLE90, Davos, Switzerland, May 1990.
8. Drake, W.O., Franz, T., Hofmann, P. and Sitek, F., "The Role of Processing Stabilizers in Recycling of Polyolefins" presented at RECYCLE91, Davos, Switzerland, May 1991.
9. Klemchuk, P.P. and Horng, P.-L., "Perspectives on the Stabilization of Hydrocarbon Polymers Against Thermo-Oxidative Degradation", *Polymer Degradation and Stability*, **1984**, 7 131-151.

RECEIVED April 6, 1992

Chapter 8

Evaluation of Mechanical Properties of Recycled Commingled Post Materials

Liqun Cao[1], R. M. Ramer[2], C. L. Beatty[1]

Department of Materials Science and Engineering, University of Florida, Gainesville, FL 32611
[2]Florida Department of Transportation, 2006 Northeast Waldo Road, Gainesville, FL 32609

As plastics are being more widely used in industry, much more plastic wastes needs to be recycled for both economical and environmental reasons. As a result, recycled plastics are becoming an economic source of engineering materials for construction and highway applications. However, the effective utilization of recycled plastics requires that the property changes that occur due to recycling be documented. Commingled post-consumer plastic scrap is currently melt-extruded into post and board shapes. The Florida Department of Transportation (FDOT) has been seeking the possible use of recycled plastics for highway applications such as guard rail posts, interstate highway boundary fence posts, sign posts, barricade sign substrates, etc.[1]. All of these applications require the study of mechanical properties since such materials must meet state specifications.

The project has been divided into the studies of the mechanical properties, paint adhesion and environmental effects, etc. In this report, the commingled post mechanical properties such as flexure strength, tensile and compressive strength, and dynamic mechanical properties will be discussed.

EXPERIMENTAL

MATERIALS

Commingled posts studied were supplied by different companies. The posts re labeled according to a pre-established code by suppliers. The labels and mensions are in Table 1.

0097–6156/92/0513–0088$07.25/0

FLEXURAL TEST

The flexural tests on recycled commingled posts were conducted on the Forny Compression Test System.

All the post samples indicate cross sections are distinctly different and contain two parts: the skin part which has relatively higher density and the core area which contains the voids. The pictures of sections from the molding end, center and the other end of post B2-2, which are labeled B2-2-a, B2-2-center, B2-2-b, respectively, are shown in Figure 1. The cross section pictures of post c-1 are shown in Figure 2. The fixture for the flexure test, a four point bending unit, is shown in Figure 3. The distance between the lower points used for supporting samples was 18", and the distance of the upper two points of the fixture used to compress samples was 6".

The lengths of the flexure test samples are 30". Post label designations are as described. The first part of the sample code indicates code number of the post and the second part of sample code indicates the sample is from either molding end portion of the post "END" or the center of the post "CENTER".

In order to study the stress-strain relation in detail, since the instrument can not print strain and displacement data, two video cameras were employed during the flexure tests of sample E-7-center, E-7-end, E-9-center and E-9-end. The tests were done at FDOT. One camera was focused on the central point of the flexural sample on the testing machine with a line scale (for deflection) mounted beside the flexed sample. The other camera was focused on the load display screen. The video images recorded during the tests were timed and then analyzed by video image analyzing software, JAVA (Jandel Corp.). Two sets of data were recorded: center point position vs time (indicated by the stop watch) and compression load vs time (also indicated by stop watch). The two stop watches were simultaneously started. The video were subsequently analyzed to get the displacement and strain data that was needed.

TENSILE TESTS

The flexural tests of recycled commingled posts provided the actual information on the mechanical capability of the posts. The flexural load to failure (i.e. the flexural strength) is indeed a structural property rather than a material property since the flexure property is not only dependent on the mechanical properties of the post material but also on the type of cross section. In order to evaluate the material properties and predict the flexural properties of the recycled commingled posts, detailed studies focused on tensile mechanical properties from remolded samples, samples from skin areas and compression properties of the samples from the core area with voids.

Table 1
The Recycled Commingled Posts

Labels	Length (in.)	Cross section area (in.^{2})	Type of cross section
B2-1	---	26.25	rectangular (3.5"x7.5")
B2-2	---	26.25	rectangular (3.5"x7.5")
C-1	95	28.27	cylinder (d = 6")
C-14	95	28.27	cylinder (d = 6")
E-7	77.5	41.25	rectangular (7.5"x5.5")
E-9	94.5	28.27	cylinder (d = 6")
BC	---	12.96	square (3.6"x3.6")

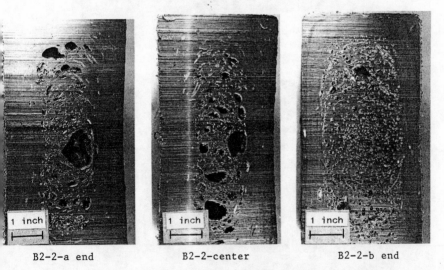

B2-2-a end   B2-2-center   B2-2-b end

Figure 1: Cross Sections of Post B2-2 molding end, center, and the other end.

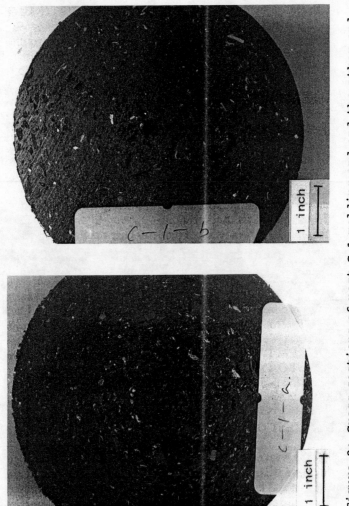

Figure 2: Cross sections of post C-1 molding end and the other end.

The tests were conducted on the MTS 880.14 Servo-hydraulic testing machine controlled by MTS 418.92 Microprofiler Controller equipped with a 1000 kg force load cell. The test data were recorded by the computer software LABTECH ACQUIRE developed by Laboratory Technologies Corporation and subjected to further analysis.

Thin recycled commingled plastic slides were cut from both the center and molding end of post C-1. The slides were further broken into small pieces, heated by platen heater to approximately 210 °C for 45 minutes, and then compressed into 5.95"x5.95"x0.13" plastic sheets via a RAM force of 3500 lb maintained for 30 minutes followed by water cooling. All the sheets were made under the same conditions and carefully labeled so that sheets from the center part of the post and the end part can be distinguished. The test samples were cut from these remolded sheets by means of a TENSIEKUT machine.

Samples c-1-c1-1, c-1c1-2, c-1c1-3, c-1-c1-4, c-1-c1-5, c-1-c2-2, c-1-c2-3, c-1-c2-4, c-1-c2-5 and c-1-c2-6 are from the center of c-1. Samples c-1-a7-1, c-1-a7-2, c-1-a7-3, c-1-a7-4, c-1-a7-5, c-1-a7-6 and c-1-a7-7 are from the molding end of post c-1. The pictures of an unbroken and a broken sample are shown in Figure 8. The samples are 5.2" long and 0.13" thick. The working sections of the samples are 2.5" long and 0.5" wide. Samples from both center and molding end of post c-1 were tested at strain rates of 0.2 in./min and 2.0 in./min at room temperature.

By examining the cross section of post E-9 near the flexural fracture surface, it can be observed that the core area shows many voids; and the high density skin is about 1" thick. However, the skin area is composed of two layers, an outside black layer about 0.5" thick, and a gray inner layer also about 0.5" thick as illustrated in Figure 4. To study the mechanical properties of the recycled commingled post, 13 samples from the black layer and 6 samples from the gray layer in an orientation parallel to the direction of the post were cut. The samples from the black layer were labeled DWBBLK1 to DWBBLK13, and the samples from the gray layer were matched DWBGRY1 to DWBGRY6. The samples were 6.45" long, 0.73" wide and 0.415" thick. The tests were conducted at the strain rate of 5 in./min at room temperature.

COMPRESSION TEST OF THE CORE AREA

The core area is the weakest region in the recycled commingled post. Eight samples were taken from the core region of post E-9 parallel to the direction of the post and labeled comp01 to comp08. A total of 17 samples were taken from the region in the direction vertical to the post direction and labeled compv01 to compv17. The samples were cylindrical with a diameter of 1.1" and length 2.15". The compression tests were also conducted at the same MTS testing system mentioned with a special compression fixture.

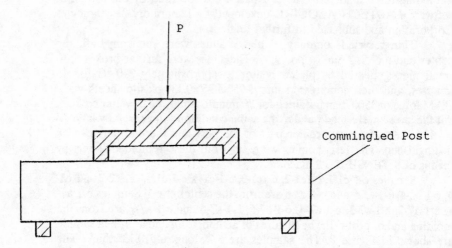

Figure 3: The fixture of flexural test of the commingled posts.

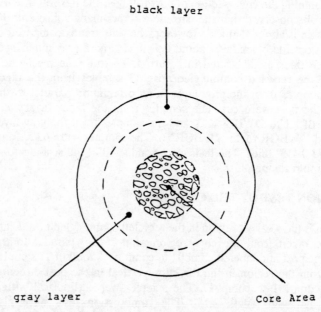

Figure 4: The structure of post E-9.

DMTA TEST

Dynamic mechanical properties were studied on the gray skin region of post E-9. Three samples were taken from the gray layer of skin parallel to the post direction and labeled E9C3P1, E9C3P2 and E9C3P3, respectively. The first two letters of the sample code E9 stands post E-9. C3 stands number 3 slide cut from the center of the post. P stands parallel direction. The other three samples were taken from the same region but vertical to the direction of the post and labeled E9C3V1, E9C3V3 and E9C3V3.

TEST RESULTS AND DISCUSSION

FLEXURE TEST RESULTS AND DISCUSSION

It can be seen that all the flexural samples are broken within the range between the two upper contact points, which indicates the failure mode was pure bending with a bending moment of

$$M = P\frac{a}{2} \qquad (1)$$

where, M = bending moment
P = compression load
a = the distance between the supporting point and the adjacent upper loading point of the fixture.
In this case, a is 6". The flexural testing results are shown in Table 2.
The load-displacement curves obtained from the two sets of data for E-7-center and E-7-end obtained by video cameras are shown in Figure 5 and in Figure 6 for E-9-center and E-9-end.
Further analysis and discussion will be detailed after a discussion of the mechanical properties of the post sample materials from the skin region and the core region.

TENSILE TEST RESULTS OF REMOLDED SAMPLES

The stress-strain curve for the sample c-1-c2-3 is shown in Figure 7. The test results are summarized in Table 3.
Statistical analysis was applied to samples from both the center and end parts with 0.2 in./min and 2 in./min strain rates. Both the strength and modulus have normal distributions. The difference between the average strength of samples from the center part and the samples from the end part

Table 2
Flexural Test Results

Sam. code	*Type	Thickness (total)	Width (total)	Thickness (core)	Width (core)
B2-1-END	R	3.5	7.5	-	-
B2-2-CENTER	R	3.5	7.5	3.5	5.5
C-l-CENTER	C	6	6	4.5	4.5
C-14-END	C	6	6	3.5	3.5
E-7-CENTER	R	5.5	7.5	3.5	5
E-7-END	R	5.5	7.5	4	5.5
E-9-CENTER	C	6	6	4	4
E-9-END	C	6	6	4.5	4.5
BC	S	3.6	3.6	2	2

Sam. code	*Type	Load (lb)	Stress (Psi)	Moment (lb x in)
B2-1-END	R	14879.16	566.8251	44637.48
B2-2-CENTER	R	21864.26	832.9241	65592.78
C-1-CENTER	C	19660.41	695.3453	58981.23
C-14-END	C	25213.62	891.7501	75640.86
E-7-CENTER	R	32572.27	789.6307	97716.81
E-7-END	R	35859.38	869.3183	107578.1
E-9-CENTER	C	26321.78	930.9433	78965.34
E-9-END	C	19062.75	674.2074	57188.25
BC	S	7134.52	550.5030	21403.56

*Type: C - Circular Cross Section
          R - Rectangular Cross Section
          S - Square

Unit of Length: in.
Unit of Load: lb.
Diameter = Thickness = Width If the Cross Section Is Circular
LENGTH OF SAMPLES: 30"

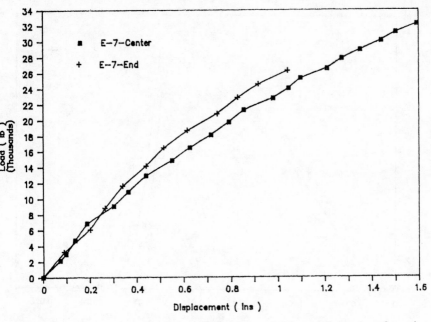

Figure 5: The load-displacement curves of flexural tests of post
E-7 center and and E-7 end.

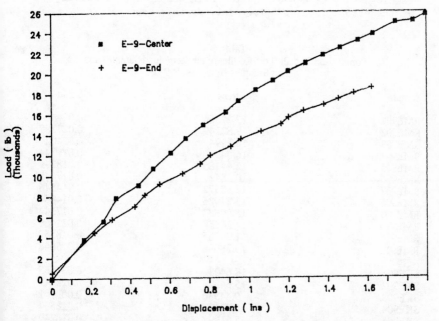

Figure 6: The load displacement curves of flexural tests of post
E-9 center and E-9 end.

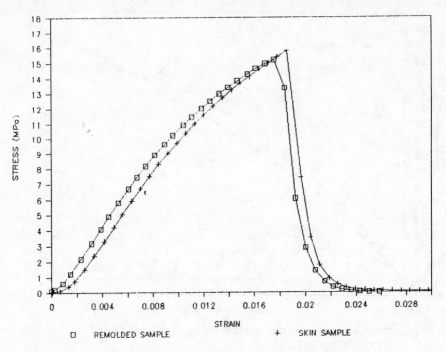

Figure 7: The comparison of the stress-strain curves of the remolded samples RME909 and the skin sample SKE904 tested at the stroke rate of 5 in./min., at room temperature.

Table 3
The Comparison Tensile Test Results of the Remolded Samples and the Skin Samples from Post E-9

Sample	Stress	Strain
RME901	17.27564	0.02486
RME902	19.82249	0.027285
RME903	15.7994	0.02402
RME904	16.549	0.022744
RME905	12.44185	0.018249
RME906	10.2842	0.017675
RME907	13.58372	0.014546
RME908	13.27926	0.014854
RME909	15.20983	0.017544
RME910	17.2657	0.021466
RME911	11.00765	0.018557
RME912	10.78698	0.015648
SKE901	18.6403	0.034583
SKE902	15.64324	0.02237
SKE903	9.44336	0.009741
SKE904	15.7943	0.018557
SKE905	15.61092	0.018249

parts are from the same sample space and have the same mechanical properties shown in Figure 8.

Combining samples from the two parts, the accumulative probability for data at 0.2 in./min is shown in Figure 9.

Both the modulus and strengths of samples tested at 2 in./min, are higher than those tested at 0.2 in./min. These results are typical in recycled noncomingled plastics.

TENSILE TEST RESULTS OF SAMPLES FROM SKIN REGION

The test results are listed in Table 4, and the stress-strain curve for samples DWBBLK4 and DWBGRY6 are shown in Figures 10 and 11.

The strength of the black layer is about 6% greater than the gray layer. However, the modulus of the black layer is about 7% lower. Both layers behave as brittle materials although the gray layer is more brittle. The comparison of the testing results of the skin samples and remolded samples will be discussed in [3].

COMPRESSIVE TEST RESULTS AND DISCUSSION

The samples described show the same mechanical behavior with the test results listed in Table 5.

The strength in the parallel direction is about 25% higher than in the vertical direction, but the modulus is about 40% lower than the vertical direction.

In comparing the moduli in the skin and core regions, the skin modulus is about seven times greater than the core region (Figure 12); and the strength is about four times that in the core region.

DMTA TEST RESULTS AND DISCUSSION

The comparison of dynamic mechanic properties in the parallel and vertical directions are shown in Figure 13 and Figure 14. There seems to be no significant difference between the properties of the two orientations.

FLEXURAL AND COMPRESSION TESTS OF SAMPLES LABELED Z

Flexural and compression tests on samples labeled Z1-3-1, Z1-3-2, Z1-3-3, Z1-4-1, Z1-4-2, Z1-4-3, Z1-5-1, Z1-5-2, Z1-5-3, Z3-3-1, Z3-3-2, Z3-4-1, Z3-4-2, Z3-5-1, Z3-5-2, Z3-6-1 and Z3-6-2 were conducted by Forney Incorporated, RT. 18, R.D. 2, WAMPUM, PA 16157. One comparison of flexural and compression load - displacement curves together with sample dimensions, and strengths for sample A1-3-3 is shown in Figure 15. The comparisons of the results shown that compressive strengths are much

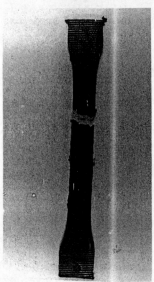

Figure 8: Photo of remolded post skin ASTM 638 tests.
top: before testing
bottom: after testing

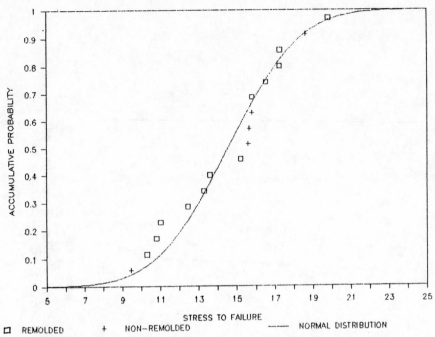

Figure 9. The accumulative probability of the stress to failures of the remolded and the skin sample from post E-9 at the stroke rate of 0.2 in./min. at room temperature.

Table 4
Test Results of Samples from the skin Region of Post E-9

Samples	Strength (KPa)	Modulus (MPa)	Strain
DWBBLK01	21846.53	1088.946	0.03363
DWBBLK02	15255.62	1297.719	0.01485
DWBBLK03	13354.73	1204.76	0.01251
DWBBLK04	14687.88	1481.991	0.01916
DWBBLK05	14638.33	1324.555	0.02307
DWBBLK06	17551.34	1365.692	0.02463
DWBBLK07	16366.32	1481.573	0.01642
DWBBLK08	18513.92	1374.362	0.01916
DWBBLK09	16242.96	1321.788	0.01838
DWBBLK10	19896.1	1236.237	0.02502
DWBBLKl1	17427.99	1480.932	0.02385
DWBBLK12	20636.74	1275.89	0.02816
AVERAGE	17296.81	1309.213	0.021444
DWBGRY01	16267.48	1522.046	0.017989
DWBGRY02	13465.70	1165.139	0.013688
DWBGRY03	16391.09	1370.091	0.009387
DWBGRY04	19908.74	1537.005	0.013295
DWBGRY05	16094.84	1429.87	0.012903
DWBGRY06	15489.94	1413.976	0.012903
AVERAGE	16269.63	1406.354	0.013360

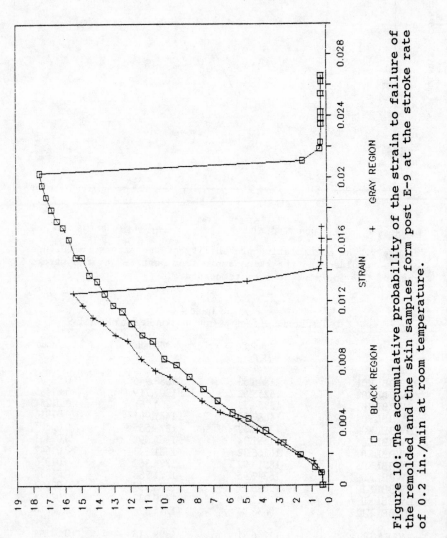

Figure 10: The accumulative probability of the strain to failure of the remolded and the skin samples form post E-9 at the stroke rate of 0.2 in./min at room temperature.

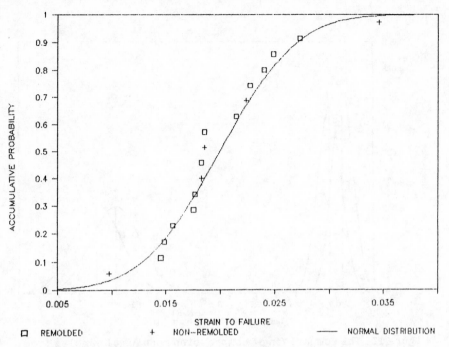

Figure 11: The comparison of stress–strain curves of sample
DWBBLK06 from the black skin region and sample DWBGRY06 from the
gray skin region of post E-9 at the strain rate of 5 in./min. at
room temperature.

Table 5
Compression Results of Samples from the
Core Region of Post E-9

Samples	Strength (MPa)	Modulus (MPa)
COMP01	6.780784	107.7606
COMP02	3.748227	276.4698
COMP03	5.089347	158.6849
COMP04	5.152745	103.9502
COMP05	4.211116	181.9306
COMP07	5.445353	116.4238
COMPV01	3.872992	134.9963
COMPV02	5.404307	174.7198
COMPV03	4.830470	117.7706
COMPV04	3.577132	201.5406
COMPV05	4.526889	305.4197
COMPV06	3.512515	279.2443
COMPV07	3.884371	280.8443
COMPV08	3.235756	258.9694
COMPV09	4.661408	274.8458
COMPV10	2.765958	254.5132
COMPVLL	2.781808	280.6655
COMPV12	2.190902	120.6260
COMPV13	3.040278	301.0342
COMPV14	3.833164	271.4699
COMPV15	3.227222	212.3330
COMPV16	2.88544	226.3387

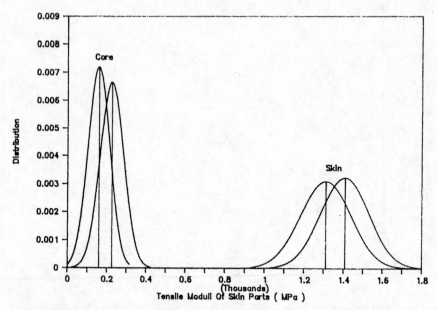

Figure 12: The comparison of compression modulus of samples from the skin region and the core region and the core region of post E-9.

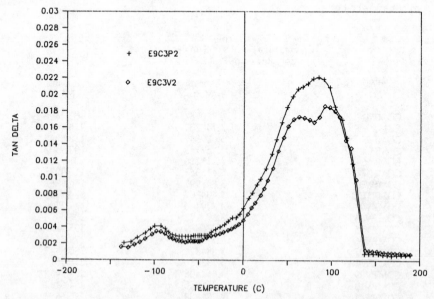

Figure 13: The comparison of tan delta of sample E9C3P2 and sample E9C3V2.

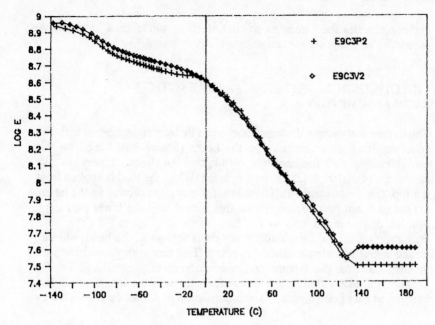

Figure 14: The comparison of Log E' of sample E9C3P2 and sample E9C3V2.

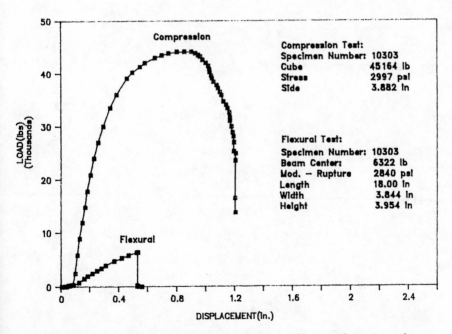

Figure 15: The load displacement curves of flexural test and compression test of sample Z1-3-3.

higher than flexural strengths so that studying alone then compressive properties of the posts are not adequate.

THE DISCUSSION OF FLEXURAL STRENGTH
BASIC ASSUMPTION

The flexure test sample discussed above can be treated as a beam with two equal loads applied vertically on the beam (Figure 16). Based on this consideration, the fundamental assumption of beam theory in the mechanics of materials [2] can be employed. When the load is applied from the top, the beam was bent. Therefore, for any cross section in the beam, the upper beam part has negative displacement; and the lower part has positive displacement due to the turn angle, θ (Figure 17). The basic assumption is that, during loading, any cross section of the beam will be turned θ, but will always remain in plane. This has been proved for pure bending such as the flexure test previously mentioned. Based on the assumption, take a small element of the beam with the length of dZ and bending angle of $d\theta$ (Figure 17), the displacement Δ, can be calculated by

$$\Delta = y \, d\theta \tag{2}$$

and therefore, the strain

$$\varepsilon = y \frac{d\theta}{dZ} \tag{3}$$

and stress

$$\sigma = E\varepsilon \tag{4}$$

$$\sigma = Ey \frac{d\theta}{dZ} \tag{5}$$

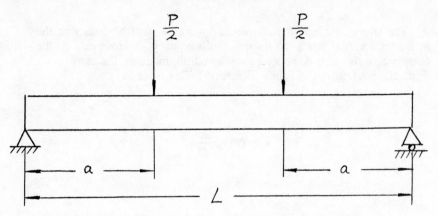

Figure 16: The structure of four point bending beam.

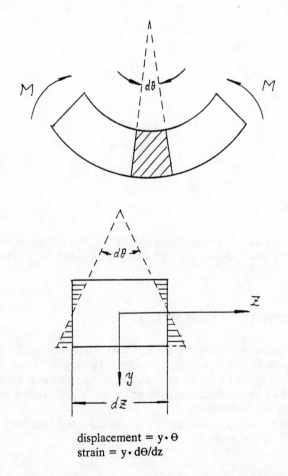

displacement = y · Θ
strain = y · dΘ/dz

Figure 17: The small beam element from the four point bending beam.

The stress distribution is shown in Figure 18. It is obvious that the maximum tensile stress is in the position of y_{max}. However, if the compressive modulus does not equal the tensile modulus, the zero displacement position will shift (Figure 19). This leads to

$$\varepsilon = (y+y_0)\frac{d\theta}{dZ} \tag{6}$$

$$\sigma = (y+y_0)E_T\frac{d\theta}{dZ} \tag{7}$$

in tension, and

$$\sigma = (y+y_0)E_c\frac{d\theta}{dZ} \tag{8}$$

in compression

where y_0 is the distance between the center of the beam cross section and the zero position and it can be determined by a force balance in the cross section.

POST B2-2 WITH RECTANGULAR SHAPE

The recycled commingled post B2-2 is made from the same recycled materials as post E-9 except that B2-2 has a rectangular cross section whereas E-9 has cylindrical across section. Before applying beam theory to the flexure test on recycled commingled posts, one simplification and one failure criterion can be utilized.

Simplification: The core area of the post does not contribute to the flexure load. First, the strain

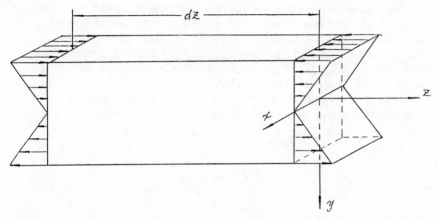

Figure 18: The tensile and compression stress on the beam element.

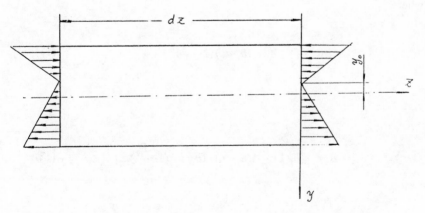

Figure 19: The zero position Yo of the beam element.

$$\varepsilon = (y + y_0) \frac{d\theta}{dZ} \tag{9}$$

is almost zero in the center of the cross section; therefore, the stress response is lower. Second, the modulus of the core area is much lower than the skin area. This also reduce the stress in core area. Therefore, the material in the core area can be neglected in this theoretical calculation.

Failure criterion: if the maximum stress in the beam reaches the strength of the material, flexural failure will occur.

In order to calculate the maximum tensile stress, y_0 must be determined first. For the rectangular post, the y_0 value can be determined by

$$y_0 = -\frac{B + \sqrt{B^2 - 4AC}}{2A} \tag{10}$$

where:

$$A = (b_1 - b_2)(1 - \frac{E_T}{E_c}) \tag{11}$$

$$B = -[b_1(1 + \frac{E_T}{E_c})(h_1 - h_2) - (b_1 - b_2)(1 - \frac{E_T}{E_c})h_2] \tag{12}$$

$$C = \frac{1}{4}(1 - \frac{E_T}{E_c})(b_1 h_1^2 - b_2 h_2^2) \tag{13}$$

The cross section parameters b , b , h and h are indicated in Figure 20. The maximum flexural load can be predicted by

$$P_{max} = \frac{2}{3}(1+\frac{E_c}{E_T})\frac{[b_1(\frac{h_1}{2}+y_0)^3 - b_2(\frac{h_2}{2}+y_0)^3]}{a(\frac{h_1}{2}+y_0)} \qquad (14)$$

where S is the tensile strength in the skin region.   For post B2-2, h1-h2=2", b1-b2=2". The average strength of both the   black skin region and gray skin region is 16.97 MPa, the ratio E_r / E_t =   0.57. The calculated y_0 value is 0.3897 in. and predicted maximum flexural   load 22303.31 lb. The value determined by actual flexural test is 21864.26   lb., only 2% different.

POST E-9 WITH CYLINDRICAL CROSS SECTION

The post with cylindrical cross section utilizes the same theory. However, the derivation of the equation for flexure samples with cylindrical cross sections is much more complicated. Considering the difficulty, y_0 is determined numerically; and P_{max} is predicted by numerical integration. The mathematical derivation are listed in Appendix A.

The calculated y_0 for E-9 is 0.45". It should be noted that, for rectangular posts, the prediction is very accurate since the load contribution regions, (the lower skin region), have the same stress value and reach their strengths simultaneously. However, for cylindrical posts, the stresses in the skin part are different. They reach the strength value gradually. This results in errors because the stress-strain curve of the skin material cannot be perfectly linear. To overcome some of the errors, it is assumed that, for cylindrical recycled plastic posts, flexure failure occur if tensile strength is reached in the whole range of the lowest skin region (Figure 21). The predicted values are 21406.5 lbs, for E-9-center and 16624.87 lbs for E-9-end . The values determined by flexural tests are 26321.78 lb and 19062.75 lbs, respectively. The respective predicted errors are 18.67% and 12.789%. This means that the whole post property can be predicted by small scale sample test results which can be easily conducted in laboratory.

It should be mentioned that, although the contribution of the core region to the flexural properties of the recycled commingled posts is very small, the core region may contribute to the impact properties since the post core can absorb some impact energy during impact tests. Recycled

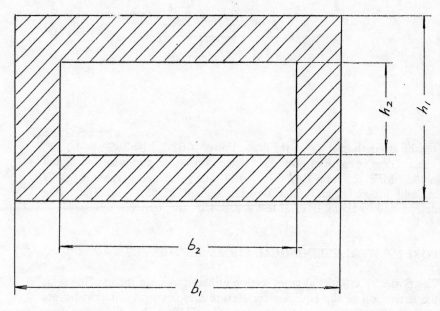

Figure 20: The dimension of cross section of rectangular beam.

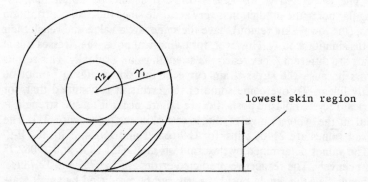

lowest skin region

Figure 21: The dimension of cross section of cylinder beam.

commingled posts composed of skin with high density and good mechanical properties and core region with very lower density materials and voids seem to be superior structures by saving weight and material while retaining relatively good mechanical properties.

SUMMARY

The flexural properties of recycled commingled posts and the mechanical properties of remolded samples and samples from the skin and core regions have been studied in detail.

The flexural properties of the posts were analyzed by beam theory. The analysis shows that the skin region is the main contribution region to the flexural properties. The core region may improve the impact properties of the posts since it can absorb some impact energy.

Dynamical mechanical tests show that there are no significant orientation effects on the mechanical properties of the post skin.

ACKNOWLEDGMENTS

The authors would like to thank Kathy Blankenship for the wordprocessing and preparation of this manuscript.

REFERENCES

1. R. Ramer S.G. Byun and C.L. Beatty, Properties of Extruded Sheet Recycled High Density Polyethylene(HDPE) Milk Bottles and Commercially Available Post Materials,SPE Recycling," RETEC, Charlette, N.C. Oct 30- 31, 1989
2. S.Timoshenko and J. Gere, Mechanics of Materials, 1972

RECEIVED August 28, 1992

Chapter 9

Accelerated Warpage and Water Sorption Tests on Recycled Plastic Posts

I. Liedermooy[1], C. L. Beatty[1], and R. M. Ramer[2]

[1]Department of Materials Science and Engineering, University of Florida, Gainesville, FL 32611
[2]Florida Department of Transportation, 2006 Northeast Waldo Road, Gainesville, FL 32609

It is the intent of Florida law #1192 that the Florida Department of Transportation (FDOT) utilize recycled plastics wherever feasible. The substitution of plastic posts for wood posts was judged to be possible due to minimal physical property requirements, compared to other FDOT applications. However, the warpage and water absorption characteristics must be comparable to or better than wood over the life of the product. In addition the greater cost of the recycled posts requires that the long term properties be documentable (where possible) to determine if the manufacturer's claimed lifetimes of 2-5 times the usable lifetime of wood is realistic. Thus, a series of laboratory tests have been developed to help screen the available recycled plastic products for more in-depth field tests. Over a long period, 5-10 years, a correlation between the accelerated laboratory tests and the field tests will be developed to predict recycled plastic post lifetimes and minimize the need for field tests of longer duration.

A. Accelerated Aging Test for Warpage

The objective of this experiment was to develop a protocol for rapid laboratory tests that predict the long term warpage characteristics of posts constructed from recycled plastic material. This test was used to determine those warpage characteristics due to residual stresses from the fabrication process. The test accelerated aging and was based on a time-temperature superposition hypothesis that has been demonstrated for the physical properties of many polymers. A primary effect of aging is a shift in the intrinsic relaxation time distribution to longer times. These distributions can be used to predict the effect of aging on mechanical properties (1).

0097–6156/92/0513–0113$06.25/0

Experimental

Because no protocol had previously been established several experiments were conducted so that the most accurate assessment of warpage characteristics could be made. The final technique implemented involved five different manufacturers' fence posts, containing at least 50% by weight recycled plastic material. The FDOT. would not allow the disclosure of the manufacturer's name therefore these posts were labeled A, B, C, D, E, and F with E being a standard wood post that is currently used. We are also prevented from analyzing the specific composition, however, all posts were polyolefin in nature. Post D contained a significant amount (approximately 50%) of sawdust and post C also contained some cellulosic derived material. Initially, posts were cut to 32 inches and the top of each post was marked by two perpendicular lines labeled A and B from which two diameter measurements were made. Each post was further marked down the side at 15 points, 2 inches apart and placed in an apparatus designed to measure any warpage induced by heat treatment. The distance from the base of the apparatus to the bottom of the post was then measured at four points about each of the fifteen marked positions. In earlier techniques hog wire was stapled in each of the fifteen sections of the post to study the effect of high temperatures on MTS staple pullout measurements. In order to facilitate these tests, the hog wire was replaced by a staple and chain. Stands were then constructed to ensure that the posts remained in a vertical position during heat treatment, thereby measuring gravity effects on warpage. Posts were then placed in the oven at either 120°C, 128°C, 125°C, or 130°C respectively. Initial measurements of length, diameter at the top of the post and circumference at the center of each post were also taken and periodically repeated throughout the experiment. Posts were removed and initial measurements were repeated before warpage persisted to the extent that the posts could no longer be placed in the apparatus.

Results

Results were represented in graphs describing the change in diameter, circumference and length as a function of time. Also examined were graphs describing the change in the dimensions at four locations along the 15 positions for each post. In all posts initial (after 3 hours of testing) increases in length and diameter were observed due to the thermal coefficient of expansion of the material. Subsequent decreases in these values after a more significant amount of time (20 hours) suggested that shrinkage may be due to the collapse of voids within the post.

Initial tests were conducted at 128°C for 6.3 hours. After testing, the percent change in the distance from the bottom of the post to the base of the apparatus was calculated. It was found that the maximum percent change for posts A, B, C, and F was between 5.0% - 7.0%. This can probably be attributed to errors in measurement taking and not to the effects of heat treatment. Post D showed a higher maximum percent change of 26%, indicating some warpage had occurred. This was verified by photographs taken before and after heating. A slight curvature and a bubbling of the surface of post D was observed. The remaining posts showed no physical signs of warpage.

Thermal aging at 125°C was conducted for 4.7 days. Posts A, B, C, D, E, and F were examined. Dimensional changes along the fifteen positions of posts A, B, C, E, and F proved to be less than 6% while post D showed as high as a 91% change. Measurements of length, diameter and circumference during the heat treatment were also taken where the change in circumference as a function of time is represented in Figure 1. Slight initial (with in the first 8 hours) increases were observed for all dimensions due again to the thermal coefficient of expansion of the material. Post D showed the most serious signs of physical warpage with a significant amount of curvature seen at the bottom 1/4 of the post. The skin of post D was also bubbled due to the vaporization of residual water during testing. Other damaging effects were noted on the skin of posts A, B and E. In all cases a drying of the surface had occurred resulting in a texture change. Posts A and E were also cracked near the staple.

Another set of posts were thermally treated at 120°C for 6 days to examine the effects of lower temperatures and long term heating on warpage. The percent change in the dimensions along the post were similar to previous tests for A and B. They were in the range of 4.0% - 6.0% while changes in Posts C, E, and F were less than 3.0%. As seen in Figure 2, where the percent change is plotted as a function of position for each of the four points around the post, post D showed changes up to 53.1%. Physical observations indicated that damage, similar to previous testing, had occurred but to a greater extent. A rust-like liquid was emitted from post A around the staple while post E was coated with a type of sap emitted from the wood. Surface cracking was also seen on post A.

Final thermal aging tests were conducted at 130°C in a series of two tests. One included the entire set of posts (A, B, C, D, E, and F), while the other consisted of three samples of post D. In the first set, conducted for 6 hours, posts A, B, C, E, F showed dimensional changes less than 6.0% whereas warpage in post D was so great that it was no longer self standing and could not be placed in the apparatus for final measurements. It was decided that a second set of tests was necessary

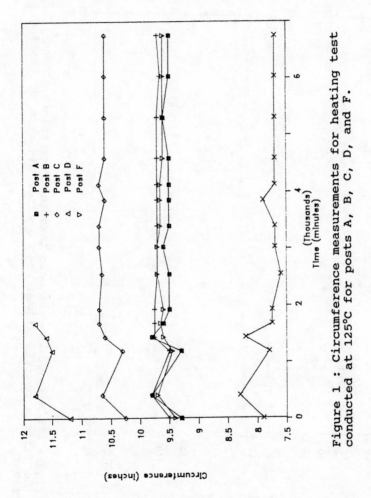

Figure 1 : Circumference measurements for heating test conducted at 125°C for posts A, B, C, D, and F.

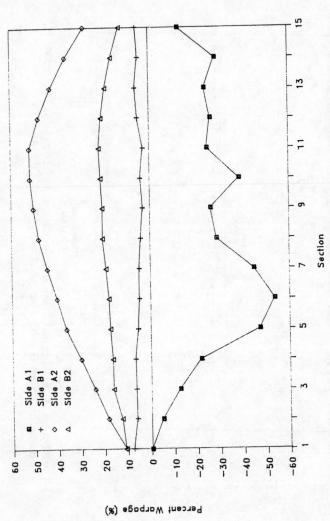

Figure 2 : Warpage measurements taken at four points along the fifteen positions down post D after six days of testing at 120°C.

because throughout experimentation, post D continually showed significant signs of warpage. In running three more tests earlier results conducted at 130°C could be confirmed or refuted. The second set was tested for 5 hours and showed a maximum 110% dimensional change along the post. Photographs of the three samples of post D after heat treatment confirmed earlier results that post D was the most susceptible to temperature induced warpage.

Summary

In conclusion, post D was the only post that showed significant signs of damage in thermal aging tests. The degree of damage was dependent on the temperature and the duration of testing. As seen in the results at 120°C a relativity long period of time (over 6 days) was necessary to achieve 53.1% changes in post D while at 130°C, 110% changes were recorded after only 5 hours of testing. All posts showed an initial increase in diameter and circumference due to the thermal coefficient of expansion of the material. Subsequent decreases, occurred as testing continued (20 hours) suggesting that perhaps internal voids had collapsed. Post D showed the most obvious signs of damage. In all cases the post was no longer self standing and a bubbling of the surface due to the vaporization of residual water. The other posts also showed some effects from the heat treatment. The surface appeared dry and in some cases, cracking was visible. Cracking was most apparent in post F (125°C), B (119°C) and post A (120°C). It can therefore be concluded that some stress relaxation was occurring in post D while the remaining posts showed no signs of residual stress but some physical damage due to the heat treatment. From these results it also appears that the main component or polymer of post D, although polyolefin in nature, may differ from the other recycled plastic posts. It appears to have a lower glass transition and melting temperature.

B. Water Absorption Tests

The objective of this test was to develop a systematic procedure for determining warpage and expansion characteristics due to boiling and salt water absorption. Diffusion properties have long been studied for virgin polymers (2), however no protocol exists for recycled plastics. Standard measurements of absorption and diffusion involve examining the change in mass as a function of time (3). Therefore the mass was recorded as a function of time. In addition dimensional changes were measured to document any damaging effects, warpage or deterioration that may have incurred during testing.

Experimental

Systematic testing procedures involving recycled plastic materials are virtually nonexistent. It was, therefore necessary to develop protocols to accurately measure accelerated water and salt water sorption characteristics. In the process three sets of tests were developed, where later tests were essentially modifications and improvements to earlier protocols. In technique 1 each of four recycled plastic posts (A, B, C, and D) were sliced into 1/8 inch thick discs. Discs were marked by two perpendicular lines, A and B, and dried in an oven at 105°C until no weight change was observed, indicating all water had been removed. Diameter measurements were taken at two positions and thickness measurements were taken at four positions around each disc. A set of discs was then placed in boiling water (100°C). Initial measurements were repeated by removing the discs and placing them on a towel to remove excess water on the surface of the samples. Approximately five minutes was allotted for the samples to cool before the weight and dimensional measurements were taken. Measurements were repeated, every 20 - 40 minutes for a total testing time of 340 minutes. Ten tests were conducted using this technique.

Technique 2 was developed as a modification of technique 1. A control or wood post, E, and another manufacturers' recycled plastic post were added to the tests. In an attempt to compensate for the lack of uniformity of some of the samples, more extensive diameter and thickness measurements were also taken and averaged. The 1/8 inch thick discs were now marked by two sets of perpendicular lines and labeled A1, A2, B1, B2, C1, C2, D1, and D2 where each line intersected the edge of the disc. As in technique 1 the initial measurements involved weighing the dried samples. Dimensional measurements, however were modified. The diameter was measured at four positions around each disc and a series of eight thickness measurements were taken at two radii. One set was measured at 1/2 inch from the center of each disc and the second set was measured at 1/2 inch from the edge of each disc. One set was placed in boiling water and another set was placed in a 5 % salt solution and boiled under pressure to maintain a constant salt concentration. Measurements were repeated every 1-2 hours in a similar manner as in technique 1. Total testing time was 600 minutes and six sets of tests were completed. Samples were also dried and weighed after the tests were completed to measure any weight loss due to deterioration of the core of the discs or weight gain due to salt absorption.

The final technique developed was technique 3. From earlier techniques it was noted that frequent measurements as well as longer testing periods were necessary in order to characterize water and salt water sorption curves. Therefore this technique was designed to measure and define functions to describe the change in weight as function of time. In both water and salt water sorption tests the weight was recorded every 10 minutes for the first 100 minutes and then repeated every 20 minutes for a total testing time of 500 minutes. Six sets of water and salt water sorption tests were completed.

Results

Results obtained using technique 1 included the weight gain, diameter, and thickness as function of time. These values were converted to percent weight gain and percent change in diameter and thickness. This normalized the data and compensated for the variation in size of the discs. Initial results showed that after 10 minutes of testing all discs exhibited, approximately, a 5% weight gain. As testing continued results for post D continued to increase while those for posts A, B and C increased only slightly, approaching a steady state condition. The percent weight gain for posts A, B and C was characterized by a logarithmic function while water sorption in post D showed a second order dependence. After 340 minutes of testing post D showed a percent weight gain of 27.6 %, approximately three times greater than values calculated for the other recycled plastic posts. Dimensional changes including the percent change in the thickness and diameter of the discs, were also calculated. From data of the percent change in diameter versus time, as seen in Figure 3, several observations were made. Post D displayed a second order dependence of the percent change in diameter with time while Posts A and C were characterized by a logarithmic function. Post B, however, showed no change in the diameter. Finally plots of percent change in thickness versus time were constructed. In all cases it was difficult to model or characterize the dependence of the percent change in thickness over time. Behavior in posts A, B and C was described by an initial increase in the thickness, within the first 60 minutes, due in part, to the thermal coefficient of expansion and perhaps initial swelling of the material by the absorbed water. The initial increase, however, fell to zero as the testing time was extended. This may be due to a loss or deterioration of the sample, most notably from the center or void region of the discs. Evidence of this was seen by a deposition of material found at the bottom of the testing apparatus after experiments were completed. Post D also displayed an increase in thickness (approximately 4.5%) and then remained at this value for the duration of testing.

Results for technique 2 were analyzed in a similar fashion as those in technique 1. Several modifications, however, were implemented. This included the introduction of a wood post, E, to act as a control in the experiment. Also introduced was another manufacturer's recycled

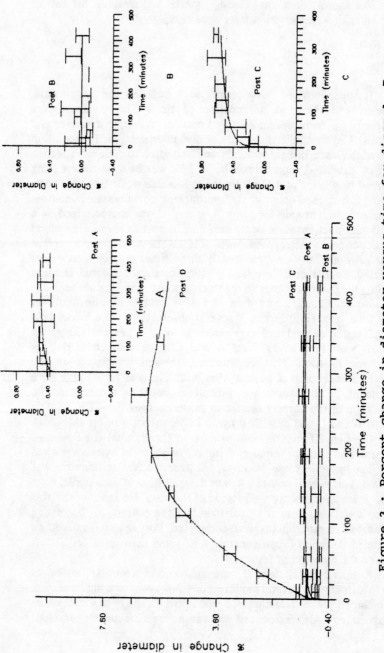

Figure 3 : Percent change in diameter versus time for discs A, B, C, and D. This graph represents the average and the standard error of the mean for ten water sorption tests using technique one. An enlargement for posts A, B, and C is represented in (A-C).

plastic post, F. Modifications to better access warpage and water sorption characteristics included additional diameter and thickness measurements. Finally the frequency that measurements were repeated was decreased but testing times were extended to 600 minutes to examine the long term water sorption.

Initial analysis, in technique 2, involved examining the changes in dimensions and weight gain for the water sorption tests. A plot of percent weight gain versus time is given in Figure 4. In all cases the most significant increase in weight occurred within the first 140 minutes. The percent weight gain then appeared to level off and reach some steady state value. In technique 2, longer testing times were implemented but measurements were not repeated as frequently. It was therefore difficult to complete any curve fitting or analysis especially since the initial portion of the curves were not accurately characterized. The only post that was characterized by an equation was post D. For the weight percent gain as a function of time the following equation was calculated;

$$\%W = 0.0564 + 0.0843 - 0.000169x^2 \qquad (1)$$

The percent change in diameter and thickness were also examined as a function of time. In both instances posts E and D showed the most significant increases. Once again due to the limited number of data points no curve analysis was completed. All posts, with the exception of post F, initially displayed a sharp increase in diameter at 140 minutes. After 140 minutes the diameter showed only a slight increase for the remaining testing time.

In technique 2 the percent change in thickness, diameter and weight gain were examined after the 600 minute testing time was completed. This was used to access maximum water sorption and dimensional changes. These values are summarized below in table I.

It is believed that in the case of posts A, B, C, and F most of the weight gain can be attributed to the entrapment of water within the voids. Upon examination of post F it appeared to contain a smaller core or void area than the other posts. This could explain why it, in turn, absorbed less water. Dimensional changes in the thickness and the diameter seemed to follow the same pattern with the wood post showing the largest percent change in diameter and thickness. An interesting observation is that although the percent weight gain in the wood post is nearly nine times greater than that of post D, the diameter changes are nearly equivalent at 4.82% and 4.33%. The remaining posts showed minimal changes with post F displaying virtually no change at all.

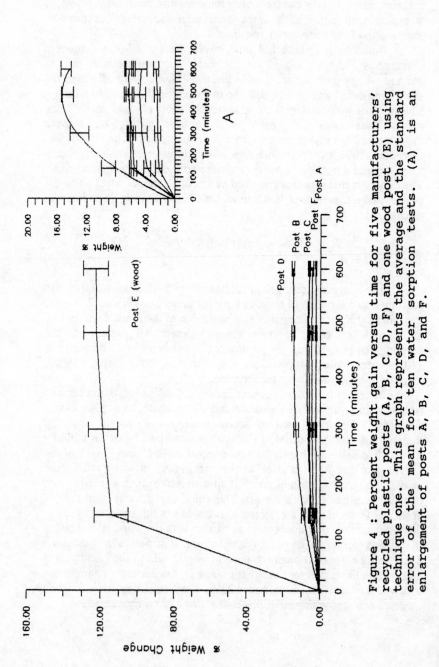

Figure 4 : Percent weight gain versus time for five manufacturers' recycled plastic posts (A, B, C, D, F) and one wood post (E) using technique one. This graph represents the average and the standard error of the mean for ten water sorption tests. (A) is an enlargement of posts A, B, C, D, and F.

The changes in the physical appearance of the samples before and after testing were also noted. Some deterioration of the core in discs A, B, C, and F was observed and residual material was found at the bottom of the testing apparatus. Post D showed the most notable signs of warpage in the form of a symmetrical curvature about the center of the disc.

The second part of technique 2 involved examining salt water sorption characteristics at 100°C. As in the water sorption tests, the most significant percent weight gain and dimensional changes occurred within the first 140 minutes, leveling off to some steady state value as testing continued. Unfortunately not enough points were sampled between time zero and 140 minutes to accurately characterize these curves showned in Table II. These results indicated that the percent weight gain of the wood post was nearly six times greater than that of post D. The percent weight gain of post D, however was nearly twice that of the remaining recycled plastic posts. Surprisingly is the degree of dimensional change seen for post D although it has a lower percent weight gain than wood it displayed larger percent changes in its diameter and outer radius. As seen above post D exhibits a 5.83% change in its diameter while post E experienced a 4.05% change. The remaining posts experienced less than a 0.5% change in their diameter. Again the percent change in the thickness at the inner radius was greatest for post D at 12.4%. Posts E and F displayed the same 6.99% increase while the remaining posts showed approximately 2% increases. Finally the percent change in the thickness at the outer radius was greatest for wood at 4.63%. Post D showed comparable dimensional changes with a 2.70% increase while unexpectedly a 3.17% increase in thickness of post A was observed. The remaining posts showed less than a 1% change in the thickness of their outer radius. It should be noted that, because of its cellulose content, the dimensional changes for posts D and E correspond to the actual swelling of the sample. Dimensional changes for the remaining posts can be probably be attributed to the deterioration or protrusion of the core material.

The physical appearance of the samples was also noted before and after salt water testing. After testing, samples appeared similar to those subjected to the boiling water tests. Damage had, however, occurred to a greater extent. Disc D showed a much greater radius of curvature while discs A, B, C and F showed a greater depletion of the core material.

Comparisons can also be drawn between results of boiling fresh and salt water tests. The same basic trends were observed with post E (wood) showing the greatest weight and dimensional changes. For all posts, in both salt and fresh water testing, the most significant weight gain and dimensional changes had occurred within the first 140 minutes of testing. Also observed was that post D showed the greatest dimensional and percent weight gain of all recycled products, which can be attributed to its cellulose content. In both types of tests post D, although absorbing less water than the wood post, displayed dimensional changes caused by swelling greater or equal to wood. Some differences, however, were observed. Posts D and E displayed greater percent weight gains in salt water testing. These values increased by 4.2% and 17.8% respectively for posts D and E. Salt water testing also accelerated the deterioration of the core and warpage in all samples.

Finally in technique 2, after water and salt water testing was completed samples were dried to remove any residual water and then reweighed. This was completed to see if any material had been lost or salt had been absorbed. Table III summarizes the average weight change upon completion of boiling and salt water testing.

In all boiling water tests final weights were lower than initial weights, indicating a loss of material had occurred during testing. Significant losses, however, were only observed for posts D and E. This justifies the appearance of the brown residue remaining in the water after testing. Results for salt water testing indicated some salt absorption had occurred in discs A, E and F while samples B, C and D showed a loss or deterioration of their material. The results of the salt water data, however, should be evaluated with caution. All discs lost weight during boiling water tests so that values of salt absorption may appear low during the simultaneous sample deterioration.

The final technique used was technique 3. This technique was developed to examine the percent weight gain as a function of time more closely. Curves were analyzed and a functional dependence of the weight gain with respect to time was developed. Figures 5 - 7 illustrate the dependence of the weight gain as a function of time. The respective equations describing the water sorption tests are represented in Table IV.

Posts D displayed a second order dependence and indicated that the diffusion was independent of concentration. The remaining posts, because of the logarithmic behavior, suggested that the diffusion was concentration dependent. This type of behavior is not uncommon in polar polymers such as wood where often a strong localization of initially adsorbed water over a limited number of sites is observed. The fraction of mobile molecules therefore increases with an increase in

TABLE I

Results from 600 minutes of fresh water testing at 100°C using technique 2

Post	Percent Weight Gain	Percent change in diameter	Percent change in thickness inner	outer
A	4.69	0.277	0.00	0.145
B	6.38	0.287	2.71	1.15
C	6.33	0.530	1.58	-0.657
D	14.8	4.82	4.49	3.52
E(wood)	122.76	4.33	4.67	5.51
F	2.62	-0.01	0.997	1.53

TABLE II

Results after 500 minutes of salt water testing at 100°C using technique 2

Post	Percent Weight Gain	Percent change in diameter	Percent change in thickness inner	outer
A	3.69	0.273	2.77	3.17
B	9.10	0.096	1.76	0.362
C	6.97	0.185	1.42	0.930
D	25.9	5.83	12.4	2.70
E (wood)	149.4	4.05	6.99	4.63
F	4.14	0.355	6.99	0.225

TABLE III

Summary of the average weight change before and after boiling and salt water testing using technique 2

Post	(F-I) weight BW (g)	(F-I) weight SW (g)
A	-0.03	0.01
B	-0.09	-0.17
C	-0.17	-0.11
D	-0.66	-0.32
E (wood)	-0.71	0.29
F	-0.04	0.05

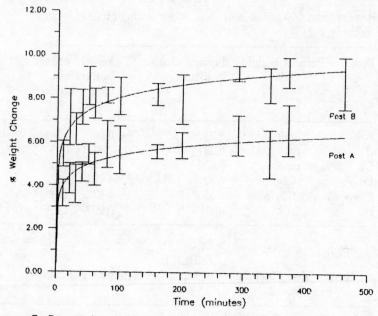

Figure 5: Percent weight gain versus time for manufacturers A and B. These graphs represent the average and the standard error of the mean for six fresh water sorption tests.

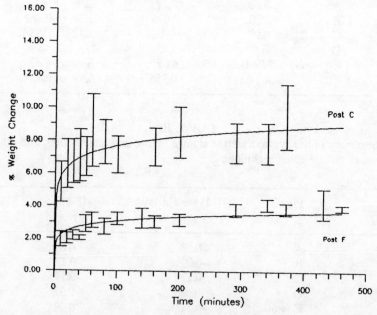

Figure 6: Percent weight gain versus time for manufacturers C and F. These graphs represent the average and the standard error of the mean for six fresh water sorption tests.

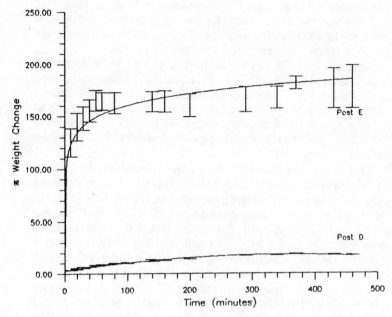

Figure 7: Percent weight gain versus time for manufacturers D and the wood post E. These graphs represent the average and the standard error of the mean for six fresh water sorption tests.

TABLE IV

Equations describing fresh water sorption at 100°C as determined by technique 3

Post	Equation
A	$\%W = 0.604\ln(t) + 2.61$
B	$\%W = 0.855\ln(t) + 4.90$
C	$\%W = 0.818\ln(t) + 3.90$
D	$\%W = 2.79 + .0869x - .000124x^2$
E (wood)	$\%W = 16.22\ln(t) + 85.63$
F	$\%W = 0.381\ln(t) + 1.32$

concentration. Studies completed on the water sorption of wood posts (4) indicated that Fick's law was obeyed and diffusion was not dependent upon concentration. These studies were, however, conducted at $20°C$ whereas the current analysis has been performed at higher temperatures to accelerate testing. Therefore the effect of temperature and the liquid/vapor concentration were not considered.

Results, after 470 minutes of testing, proved to be similar to those in techniques 1 and 2. The wood post, E, showed the largest percent weight gain at 177.07%. Comparison among the recycled plastic posts showed that post D displayed the greatest percent weight gain at 17.39%. Posts A, B and C showed comparable percent weight gains of 6.60%, 8.75%, and 9.49%. Finally post F experienced the smallest percent weight gain at 4.04%

Similar experiments were conducted to analyze the percent weight gain for tests conducted in salt water. Figures 8 - 10 illustrate the dependence of the percent weight gain with time. From these curves it can be seen that the dependence for posts A, B, C, E, and F was not logarithmic. Initially, within the first 10 minutes, a sharp increase was observed which reached a maximum and then appeared to fluctuate about some average value, behaving as a higher order polynomial. Therefore the total behavior can neither be described by a logarithmic function nor a higher order polynomial. A differential equation may have to be derived to describe these functions. There are several explanations as to why this type of behavior was observed. One possibility is that, throughout testing salt may have diffused in and out of the sample depending on the concentration gradients at that particular time and place within the testing apparatus. Hence a absorption/desorption is occurring about the saturation point. Also possible, is simultaneous salt absorption and material deterioration causing the weight to fluctuate. The percent weight gain of post D showed a much less complicated dependence with time. The second order dependence postulated in technique 2, however, was not observed. Instead the dependence was third order described by the following equation;

$$W\% = 2.43 + .119x - .000385x^2 + (4.81*10^7)x^3 \qquad (2)$$

Summary

In conclusion, the samples most effected by boiling and salt water absorption were discs D and E. From technique 2 after 600 minutes of water sorption testing the wood post, E, showed a 122.76% weight gain. This was nearly nine times greater than the percent weight gain for any

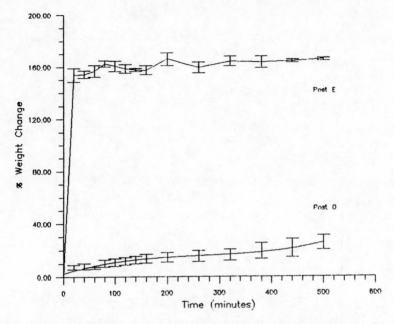

Figure 8: Percent weight gain versus time for manufacturers D and the wood post E. These graphs represent the average and the standard error of the mean for six salt water sorption tests.

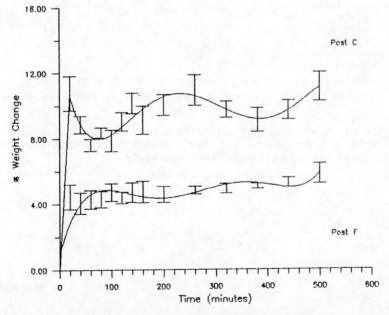

Figure 9: Percent weight gain versus time for manufacturers C and F. These graphs represent the average and the standard error of the mean for six salt water sorption tests.

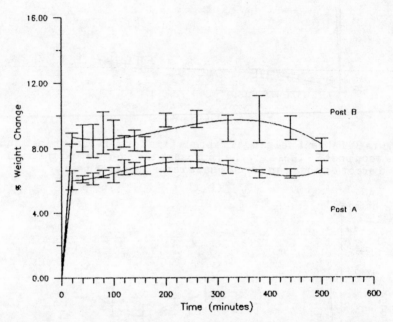

Figure 10: Percent weight gain versus time for manufacturers A and B. These graphs represent the average and the standard error of the mean for six salt water sorption tests.

of the recycled plastic posts. Among the recycled plastic posts, post D exhibited the greatest percent weight gain in all three techniques employed presumably due to its relatively high wood content. In technique 2 it showed a 14.8% weight gain. Posts A, B and C displayed gains of 4.69%, 6.38%, and 6.33% while post F exhibited the smallest percent weight gain of 2.62%. Dimensional changes in the thickness and diameter followed the same type of trend with posts D and E showing the only significant amount of change. Water sorption is known to occur via three mechanisms, where the water may be bound, or free in the liquid or vapor form. Posts A, B, and F are basically polyolefin in nature therefore the effects of bound water sorption are negligible and the percent weight gain was attributed to the entrapment of liquid and vapor within the voids. Therefore since bound water induces swelling little or no dimensional changes were observed for these posts. An interesting observation is that although the percent weight gain in the wood post, E was nine times greater than that of post D the dimensional changes or swelling were nearly equivalent.

In salt water testing similar trends were observed with the wood post, E, showing a weight percent gain approximately six times greater than any of the recycled plastic posts. From technique 2, a 149.4% percent weight gain was observed for post E while a 25.9% gain was observed for post D. Posts A, B, C and F showed percent weight gains of 3.69%, 9.10%, 6.97%, and 4.14%. In all cases the percent weight gain in salt water testing increased over values observed in boiling water testing. Most dramatically was post D which doubled its value.

Curve fitting was conducted in technique 3. It was shown that the percent weight gain, in fresh water sorption, for posts A, B, C, E, and F was characterized by a logarithmic function. Post D followed a second order dependence, observable in all three techniques. The behavior of the percent weight gain during salt water testing was more difficult to characterize. Posts A, B, C, E, and F displayed a sharp increase, of the percent weight gain, during the first 10 minutes of testing. The behavior was then illustrated by a fluctuation about some average value. This behavior was defined by a high order polynomial. The whole behavior, however, could not be defined and it is speculated that the derivation of a differential equation will be necessary.

Finally the physical appearance of the samples during testing was noted. Post D showed the most obvious signs of warpage. During testing, in both boiling and salt water, the center of the disc symmetrically protruded outward taking the shape of a shallow bowel. Also no during testing was the deposition of material at the bottom of the testing apparatus. Depletion of the core material in posts A, B, C, and F was noted in both tests but occurred to greater extent in salt water testing.

In response to this the discs were dried and weighed after testing. All discs tested in boiling water indicated a loss in mass or deterioration of the samples had occurred. In salt water testing posts B, C, and D experienced a decrease in mass while the remaining posts showed an increase. This indicated that salt absorption had occurred in posts A, E and F. The data for posts B, C and D was examined with caution because sample deterioration may have offset the mass gain due to salt absorption.

References

1.) Beatty, C., L., Harmon, J., P., "Thermal Stresses and Physical Aging," Engineered Materials Handbook, Vol 2, Engineering Plastics, p.751, ASM International (1988).

2.) Crank, J., Park, G., S., Diffusion in Polymers, Academic Press, New York, 1968.

3.) Wypych, J., Weathering Handbook, Chemtech Publishing, Toronto, 1990.

4.) Kouali, M., Vergnaud, J. Wood Sci. Technol.,(25),1991 pp.327-339.

RECEIVED August 28, 1992

Chapter 10

Phosphite Stabilizers in Postconsumer Recycling

Suzanne Dietz

GE Specialty Chemicals, P.O. Box 1868, Route 2 North, Parkersburg, WV 26102

Certain phosphite stablizers have been shown to be effective in maintaining stability of post consumer plastics during simulated processing conditions. Test methods include exposing material to heat and shear in a Haake torque rheometer and multiple pass extrusions with measurement of melt index. PET, HDPE homo and copolymer, PP and PC have been evaluated. Without sufficient stabilizer, degradation which occurs during reprocessing may render the materials unsuitable for high value products. Color is noticeably affected and molecular weight changes as measured by changes in torque or viscosity occur. Molecular weight can be stabilized through the use of phosphites and color changes minimized. Phosphites have long been known to be effective processing stabilizers and their benefit clearly extends to processing of recycled polymers.

It is no secret that the United States, and in fact the entire world, faces a solid waste problem which is approaching crisis proportions in some areas. The amount of plastic material in the waste stream is small but increasing. In a world of shrinking natural resources and dwindling energy supplies, plastic waste materials should be viewed as a potential source of precious raw materials. Industrial producers of plastic materials, recognizing the value in once processed thermoplastics, have 'recycled' plastics in large quantities for many years. A relatively recent development is post consumer recycling (reusing

0097–6156/92/0513–0134$06.00/0
© 1992 American Chemical Society

material which passes through the household or work place on its way to additional uses). The old saying "one man's garbage is another man's gold" was never more true than when the garbage is plastic.

If we consider a recycling pyramid,

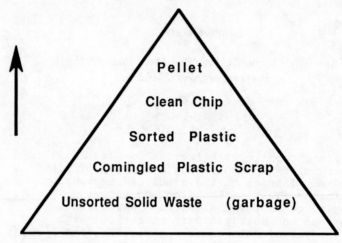

(Reproduced with permission from ref.10. Copyright 1990 Maack Business Services)

we see that each step upward increases the value of the material. As material moves up the pyramid, it becomes cleaner and has more consistent properties which allows it to be marketed into a wider range of applications. At the upper limit, recycled pellets would be equivalent in properties and appearance to virgin pellets and would be able to command comparable prices. In fact, in a society where purchase of recycled goods is mandated or socially desirable, the price of good quality recycled pellets may exceed that of virgin material. Shortages of energy and/or raw materials will push up the price of virgin polymers while mandatory recycling legislation and curbside collection systems should make polymer solid waste more available. The economics of recycling can only improve. The remaining questions are how soon, how much and how fast.

Obviously, each business involved in recycling will want to add value, and therefore margin, to their product by stepping up the pyramid. This paper describes how the use of additives (phosphite stabilizers in particular) can improve post consumer PET, HDPE, PP and PC recycle. Very small amounts of stabilizer can provide dramatic improvements in polymer properties and may enable recyclers to meet critical market demands. Principal

benefits expected from the use of phosphite stabilizers
are consistency of melt flow through multiple processing
steps and improved color retention. This should give the
material consistency during re-working and allow for an
increased percentage of recycle in virgin/recycle blends.
The proper use of additives can make most polymers
(especially recycled ones) better.

PET Recycle

Introduction. Producers of polyethylene terephthalate
(PET) have enjoyed a remarkable growth in the consumption
of PET over the past twenty five years. Most of the
early demand was for fiber applications but recent growth
has been strongest in non-fiber applications such as
film, bottles and microwaveable ovenware. (1)

   PET is an excellent polymer for packaging and its
use in food packaging widespread. Many household
articles made from PET find their way into the solid
waste stream or are discarded in careless ways.
Consequently bottle deposit legislation in populated
states has allowed PET to become the most significant
post consumer plastic to be recycled. Current recycling
efforts in PET are estimated to be 215 million pounds
annually which represents approximately 28% of the annual
beverage bottle market. (2) Proper stabilization of PET
during the reprocessing steps can prevent molecular
weight loss and maintain good color allowing for high
quality second use applications.

Degradation Chemistry. The thermal degradation of PET
has been studied extensively by Zimmerman and others.(3)
Increasing carboxyl end groups, acetaldehyde production
and color as well as decreasing viscosity are all ways to
measure the extent of the degradation of PET.

Stabilization. Processing PET in a torque rheometer
has been used in our laboratory as an accelerated method
to follow the degradation of PET. This test simulates
the multiple processing steps which would be seen in
recycling PET to second and third use products. The
details of the method have been previously published.(4)
   PET responds very well to stabilization by the
phosphite bis(2,4-di-t-butylphenyl) pentaerythritol
diphosphite (P1) as seen in Table I.

Table I. Stabilization of PET by the Use of Phosphite P1

The PET is exposed to heat and shear in a torque rheometer at 280 DEG. C and 60 rpm.

	Minutes Exposure				
	10	20	30	40	50

Intrinsic Viscosity (i.v.)

PET alone	0.69	0.66	0.63	0.60	0.56
PET + 0.1 P1	0.79	0.74	0.72	0.66	0.64

Torque - Meter Grams

PET alone	400	280	220	180	140
PET + 0.1 P1	560	540	510	470	420

Color - Yellowness Index

PET alone	25.4	29.0	38.0	45.2	46.9
PET + 0.1 P1	16.0	17.9	19.3	25.6	27.2

Free Carboxyl Ends (meq/mg)

PET alone	32.0	37.0	41.5	45.8	57.7
PET + 0.1 P1	25.0	28.0	32.0	38.0	43.8

PET with and without P1 was exposed to heat and shear in a torque rheometer at 280°C for 50 minutes. Molecular weight loss is minimized as evidenced by torque and viscosity data. There is less development of color and carboxyl groups with the stabilizer. Reduction of acetaldehyde is also achieved through the use of phosphite stabilizers. PET with 0.2% P1 or P2 bis(2,4,6 -tri-t-butylphenyl) pentaerythritol diphosphite contained half the amount of acetaldehyde at flux as samples processed without phosphite.

As expected, actual post consumer recycle (PCR) PET bottles also react favorably to phosphite use. Figure 1 shows the result of increasing stabilizer levels.

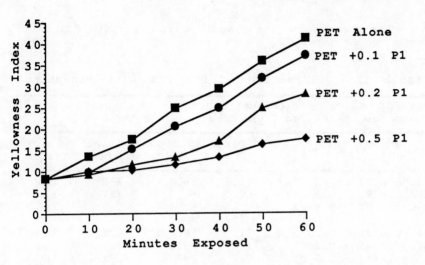

Figure 1
Color Response of PCR PET to Phosphite Stabilizer
Reproduced with permission from ref.10. Copyright 1990
Maack Business Services

The response of recycled PET to stabilizers is similar to
that of virgin material and with proper selection of
stabilizer amounts, performance properties similar to
virgin material can be achieved. Figure 2 compares
virgin bottle material to stabilized recycle.

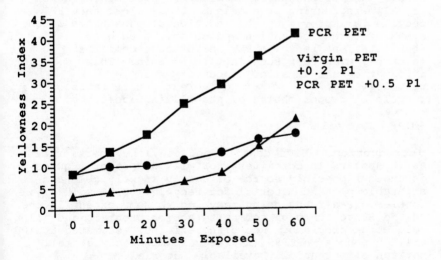

Figure 2
Comparison of Stabilized Virgin and PCR PET
Reproduced with permission from ref.10. Copyright 1990
Maack Business Services.

The results for three additional PET recycle streams are given in Table II.

Table II.Stabilization of PCR PET with Phosphites

The PET was exposed to heat and shear in a torque rheometer at 280 Deg C and 60 rpm.

Sample	Initial YI	40 Min YI	Initial Torque	40 Min Torque
Recycle #1 Green	61.9	66.7	230	115
Recycle #1 + 0.5 P1	60.1	61.0	251	140
Recycle #1 + 0.5 P2	58.9	60.4	260	186
Recycle #2 Clear	7.3	18.7	285	160
Recycle #2 + 0.2 P1	7.0	13.9	285	168
Recycle #2 + 0.5 P1	6.7	9.8	282	180
Recycle #2 + 0.2 P2	7.7	14.3	285	176
Recycle #2 + 0.5 P2	8.1	12.0	322	220
Recycle #3 Green	43.1	51.6	256	126
Recycle #3 + 0.5 P1	41.7	42.8	243	172
Recycle #3 + 0.5 P2	41.9	44.0	253	191

In the green materials initial color is quite high. However use of phosphite stabilizers prevents darkening or yellowing during processing. It is clear that PET recycle is improved by the addition of phosphites in the areas of torque retention and color. With good stabilization, recycled PET performance compares favorably with that seen in virgin resin. This information should allow processors using a recycle/virgin mix of PET to increase their ratio of recycle by proper choice of stabilizing additives.

HDPE Recycle

Introduction. High density polyethylene, particularly as it appears in containers for milk and other liquids, is another possible source of easily identifiable and separable plastic material for recycle. HDPE milk bottles are already being recycled to some extent in the United States through community action programs and pilot collection programs. Recycling efforts, estimated at 100 million pounds in 1989 (5), are claiming only a small portion of potentially available material.

Degradation Chemistry. Changes which occur in HDPE during processing are complex and can involve both cross linking and scission. Previous authors (6, 7) describe

some of these changes and factors which can affect them.
Changes of melt flow in either direction create
difficulties in predicting necessary processing
conditions as well as final physical properties.
Additives that preserve the melt flow properties of
incoming material during pelletization and/or processing
would be beneficial.

Stabilization. Phosphites are known to be effective
stabilizers of HDPE during processing. Table III shows
the effect of the phosphites P1 and tris(2,4-di-t
-butylphenyl) phosphite P3 on the melt index and color of
a virgin high molecular weight HDPE resin.

Table III. Stabilization of HDPE by the Use of Phosphites

The HDPE was exposed to multiple passes through an
extruder at 250 Deg C.

	Extrusion pass			Extrusion pass	
	1st	3rd	5th	1st	5th
	Melt index (I_{10})			Color (YI)	
Base	.71	.48	.49	1.81	9.04
Base + 500 ppm P1	.82	.81	.80	− 3.04	−0.46
Base + 1000 ppm P3	.85	.85	.86	− 0.18	11.53

Base = HDPE + 500 ppm AO1 tetrakis[methylene(3,5-di
-t-butyl-4-hydroxyhydrocinnamate)]methane

During the test, HDPE was exposed to multiple passes
through an extruder. This is another test method which
simulates the stresses a material would see during
recycling. Use of P1 insures consistency of melt flow
through 5 passes. In virgin HDPE, color development is
also retarded through use of the stabilizer.
     Post consumer HDPE homopolymer milk bottles respond
favorably to the addition of stabilizers. Figure 3
indicates the consistency of melt flow achieved with
stabilizers versus the changes in melt flow of the HDPE
which is not stabilized.

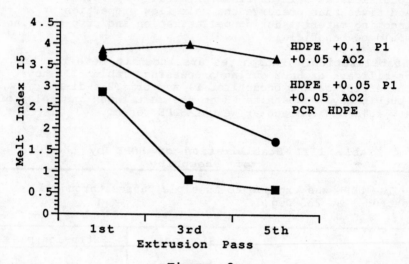

Figure 3
Response of PCR HDPE Homopolymer to Stabilization
Reproduced with permission from ref.10. Copyright 1990
Maack Business Services

This sample of recycled HDPE contained small
particles of colored caps (probably polypropylene).
Pressed plaques of the material were gray and no
differences in yellowness index could be detected.
    Another experiment exposing recycled HDPE flake to
heat and shear in a torque rheometer at 210°C indicates
that the material experiences a gradual loss in torque
from flux to 27 minutes. Torque then begins to increase
and continues increasing for 30 minutes. At this point,
the HDPE is cross linked to an extent which would make it
difficult to process and the experiment was stopped.
With 0.1% by weight of P1, the torque is constant from
flux to 60 minutes at which point the experiment was
terminated. There was no effort made to exclude oxygen
from this system.

HDPE Copolymer. Post consumer HDPE copolymer, such as
that found in detergent bottles, also responds well to
phosphite stabilizers. Figure 4 shows the effect of P1
at two different loading levels.

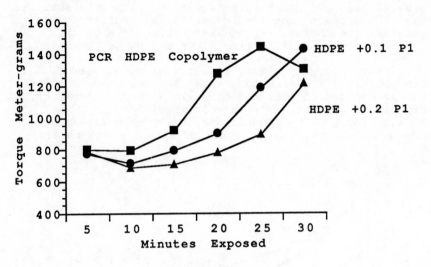

Figure 4
Response of PCR HDPE Copolymer to Stabilization

PP Recycle

Introduction. Polypropylene fiber scrap has been recycled within the industry for many years (8). However, because of limited supply, post consumer polypropylene materials have not been recycled in large quantities. Polypropylene battery cases are a notable exception, with recycled material being successfully incorporated into new battery cases. Some polypropylene scrap is also being made into filled PP for lawn and garden furniture.

Degradation Chemistry. Polypropylene degrades via a thermal oxidative free radical mechanism. The primary result is random chain scission (9). The resultant molecular weight loss is evidenced by an increase in melt flow and lowered physical properties. Without sufficient stabilizer, polypropylene cannot be successfully processed as a virgin resin. With proper stabilization, PP can be not only processed initially but also reprocessed or recycled several times maintaining consistent melt flow and consistent properties.

Stabilization. Polypropylene responds well to stabilization and normally contains a comprehensive stabilization package in its original formulation. Phosphites and hindered phenolic antioxidants are effective in maintaining melt flow in virgin PP. After

several passes through an extruder, PP without stabilizer
exhibits a sharp increase in melt flow, indicating that
considerable chain scission has occurred. With the
addition of proper stabilizers, melt flow can be held
flat.
 Post consumer PP also benefits from the use of
stabilizers. Figure 5 indicates that recycled PP battery
cases, exposed to heat and shear in a torque rheometer,
experience less torque loss when stabilized.
P1 also improves the consistency of PP melt flow after
five passes through an extruder as shown in Figure 6.

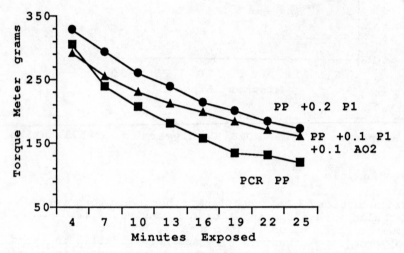

Figure 5
Torque Response of PCR PP to Stabilization

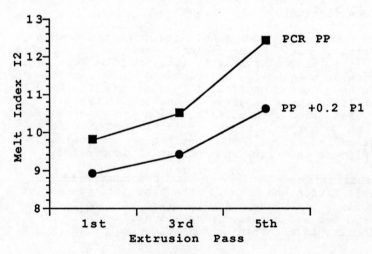

Figure 6
Phosphite Stabilization of PCR Filled PP

PC Recycle

Introduction. Polycarbonate water bottles can be re-used many times but eventually enter the waste stream where they can also be recycled. Because polycarbonate bottles are not single use items, they are not available in quantities which are usually required to make recycling attractive. However, the high price of PC combined with its excellent properties insures that a considerable amount of post consumer material is being recycled. Compact disks are another high volume consumer item which can be recycled after use.

Degradation and Stabilization. Polycarbonate is an excellent material which holds its properties well through processing. However, experiments conducted in water bottle scrap indicate that P1 can assist in maintaining color and viscosity in PC exposed to heat and shear in a torque rheometer. Figure 7 shows the torque curves of PC with and without P1 stabilizer.

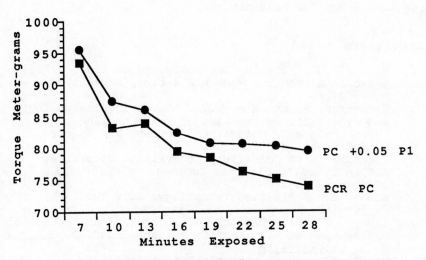

Figure 7
Torque Response of PCR PC to Phosphite
Stabilization

Conclusion. A responsible approach to the solid waste crisis must involve a multi-faceted program including source reduction, recycling when possible and waste to energy incineration. Both public policy and economics dictate that the valuable raw materials in plastic ought to be salvaged. With proper stabilization of the initial polymer and the correct additives in the melt/recycle processing, most thermoplastic materials should be able to be recycled to high quality second and third uses.

Upgrading recycled resins through proper stabilization may be the key to higher value products and better economic return for recyclers. Certain phosphite stabilizers have been shown to be effective in maintaining the melt viscosity of post consumer PET, HDPE, PP and PC during simulated processing conditions. Use of these and other stabilizers should allow valuable products to be made from plastic materials that now end up in landfills after their first use. The use of stabilizer technology to improve cleaned single polymer chip is especially attractive. The resulting compounded pellets will be able to compete with virgin pellets for many applications. Improved prices for recycled plastic pellets will help all businesses which are involved in recycling.

The author would like to acknowledge the excellent work of R. Thompson on the PET, PP and PC experiments and for the rheometer study of HDPE and the contributions of J. Stull for the extruder work on HDPE. The author would also like to thank B. Anderson for her assistance in preparing the manuscript for publication.

Literature Cited

1. Block, D.G., Plastics Technology, June 1991, p.180.
2. Leaversuch, Robert, Modern Plastics, March 1987, p. 44-47.
3. Zimmerman, H. and Kim, N.T., "Investigation on Thermal and Hydrolytic Degradation of Poly(ethylene terephthalate)", Polymer Engineering and Science, p. 680-683, mid-July 1980.
4. Fleenor, C. "A Quick Test to Evaluate Stabilizer Action in PET", Plastics Engineering, February 1986, p. 37-39.
5. Schut, J.H., Plastics Technology, July 1990, P. 109-119.
6. Rideal, G.R. and Padget, J.C. "The Thermal Mechanical Degradation of High Density Polyethylene", J. Polymer Sci., Polymer Symp. No. 57, p. 1-15, 1976.
7. Mitterhofer, F. "Processing Stability of Polyolefins", Poym. Eng. Sci., 20:692-695, 1980.
8. Bash, David P. "Recycling of Polypropylene Fiber Waste for Textile and Plastic Applications", Symposium, Polypropylene and Polyolefins in Textiles, Atlanta, Ga. March 17, 1988.
9. Davis, Thomas E., Tobias, Robert L. and Peterli, Elizabeth B., "Thermal Degradation of Polypropylene", J. Polym. Sci. Vol. 56, pp.485-499, 1962.
10. Dietz, Suzanne B., "The Use and Market Economics of Phosphite Stabilizers in Post Consumer Recycle", Recycle '90 Forum and Exposition Proceedings, Davos, Switzerland, P. 283, May 30, 1990.

RECEIVED August 10, 1992

Structures

AO1 = tetrakis[methylene(3,5-di-t-butyl-4-
hydroxyhydrocinnamate)] methane

AO2 = octadecyl 3,5-di-t-butyl-4-hydoxyhydrocinnamate

P1 = bis(2,4-di-t-butylphenyl) pentaerythritol diphosphite

P2 = distearyl pentaerythritol diphosphite

P3 = tris(2,4-di-t-butylphenyl) phosphite

Chapter 11

Selected Aspects of Poly(ethylene terephthalate) Solution Behavior

Application to a Selective Dissolution Process for the Separation of Mixed Plastics

Leland M. Vane[1] and Ferdinand Rodriguez

School of Chemical Engineering, Olin Hall, Cornell University, Ithaca, NY 14853

The solution crystallization kinetics and crystal dissolution behavior of a commercial grade of poly(ethylene terephthalate) in N-methyl-2-pyrrolidinone were characterized using turbidimetric and dilatometric methods. The importance of this information to a selective dissolution process for the separation of poly(ethylene terephthalate) from a 2–liter bottle waste was investigated. A relationship between the photosignal response data from the turbidimeter and the dilatometer extent-of-transformation data was developed which allows for the prediction of the crystallization rate constant using the more easily obtained turbidimeter data.

In the United States and, indeed, the world, a ground swell of interest in the recovery and reuse of all post consumer materials has occurred. This public concern has motivated research into processes to separate mixed post-consumer plastics from the overall waste stream as well as from each other. Mixed plastics command a relatively low price compared to virgin resins primarily due to diminished physical properties resulting from polymer–polymer incompatibility, discoloration, and degradation. In addition, the use of mixed plastics in downgrading operations is not truly "recycling" since the material does not replace virgin resins.

Using the closure of the recycling loop as a goal, interest has resurfaced in processes to separate mixed post-consumer plastics. Common methods for separating mixed plastics include air classification, hydrocycloning, flotation-sedimentation, depolymerization-purification-repolymerization, and selective dissolution. Of these, only selective dissolution and depolymerization are capable of removing bound impurities from the plastic as well as differentiating plastics based on the chemical properties of the individual polymers. These two methods should yield polymers suitable for reuse in original applications, although both suffer from increased expenses due to the complexity of equipment and higher energy requirements.

Selective Dissolution

The concept of using solvents to dissolve individual polymers selectively from a

[1]Current address: U.S. Environmental Protection Agency, Cincinnati, OH 45268

mixture is by no means new, but it is certainly receiving renewed attention. Much of the original work on solvent processes occurred in the 1970's. In one of the original research efforts in solvent processes, Sperber and Rosen (1,2) separated a mixture of polystyrene (PS), poly(vinyl chloride) (PVC), high–density polyethylene (HDPE), low–density polyethylene (LDPE), and polypropylene (PP) into three separate phases using a blend of xylene and cyclohexanone. At about the same time, Seymour and Stahl (3) were studying the use of toluene and methanol at various temperatures in a branched scheme to selectively remove individual polymers from a mixture of PS, PVC, LDPE, poly(methyl methacrylate) (PMMA), and poly(vinyl acetate) (PVAC). Other researchers concentrated on the solvent recovery of individual polymers such as polypropylene (4). In addition, numerous domestic and foreign patents were granted in the 1970's for the solvent separation and purification of thermoplastic polymers (5–9). Interest in plastics recycling diminished in the late 1970's and early 1980's as natural resource concerns lessened and the public settled into carefree throwaway habits. The solid waste problems of the late 1980's and the volatility of the oil export regions has aroused the public and brought new life to plastics separations research (10-19). Research on solvent processes has been a part of this resurgence (10-14).

Inclusion of PET in Mixed Waste. Early research on the solvent separation of mixed plastics did not include PET in the mixture because this versatile and valuable polymer had not penetrated the packaging market. In addition, many of the solvents chosen for use in these processes were not adequately screened for their effects on health and the environment. It was, therefore, necessary to reconsider the concept of selective dissolution with the recovery of PET and the selection of more "friendly" solvents included as research goals. Recently, researchers at Rensselaer Polytechnic Institute have studied the separation of PVC, PS, LDPE, PP, HDPE, and PET using either xylene or tetrahydrofuran (12). The polymers are dissolved in batch mode with the separation based on the solvent temperature which controls the dissolution rate of each resin. Once dissolved, the polymer solution is exposed to elevated temperatures and pressures before the solvent is flash devolatilized. This process suffers from the same limitations which beset all single solvent systems. When only one solvent is used to dissolve a wide range of polymers, the selectivity is significantly decreased due to the unintended partial dissolution of polymer "A" when polymer "B" is the target polymer and also carryover of undissolved "B" when "A" becomes the target polymer. Relying solely on the temperature dependent dissolution rate of polydisperse and semi-crystalline polymers significantly reduces the purification capacity of the process. A considerable improvement in selectivity can be achieved by using multiple solvents which are compatible with only a limited number of polymers. In this way, polymer dissolution is not just a function of the solvent temperature, but of the polymer-solvent interaction as well.

Combining Separation Technologies. An even more advantageous separation scheme is the combination of low-cost sink-float separation technology with a multi–solvent selective dissolution process. A sink-float process would serve as the first separation stage for shredded plastic waste, achieving segregation into two or more streams based on the densities of the materials as shown in Figure 1. The streams from the sink-float process are then further purified using solvent-processing trains with the solvents optimized based on the compositions of the sink-float product streams. A general flow sheet of a solution process is shown in Figure 2. The first stage of the solvent process involves the removal of soluble impurities from the stream using the process solvent at a slightly elevated temperature. This temperature should be kept low enough to prevent dissolution of the target polymer. The impurities to be removed in the Solvent Washing stage are materials such as adhesives, coatings, and any soluble non-target polymer. The remaining undissolved materials are then exposed

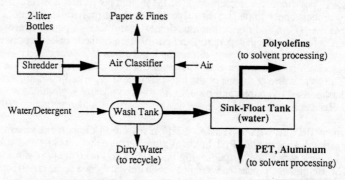

Figure 1. Schematic diagram of sink-float process for 2-liter bottles.

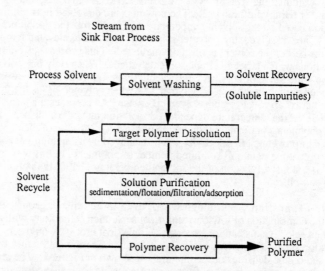

Figure 2. Simplified flow diagram for a solution process.

to the process solvent at a higher temperature than in the first stage, causing rapid dissolution of the target polymer, but leaving the remaining polymers undissolved. As the target polymer dissolves, impurities such as fillers, catalysts, pigments, and plasticizers which may be bound in the polymer matrix are released. These materials as well as the undissolved remnants of the original feed stream must be removed before the target polymer is recovered from solution. This purification can be achieved utilizing a variety of separation technologies including sedimentation, flotation, filtration, adsorption, and ion exchange. Insoluble impurities such as metals, thermosets, rocks, etc. are easily removed by filtration from the low viscosity polymer solution. This is not true for the filtration of these same materials from a polymer melt. The selective dissolution process, therefore, has a significant advantage in the area of metals removal.

The final stage in the solvent process is the recovery of the target polymer from the solution. This recovery can be achieved in many ways including temperature quenching of the solution resulting in polymer crystallization or precipitation, shock–precipitation by addition of a non-solvent, or flash devolatilization (polymer deposition). Addition of a non-solvent appears to be the most attractive because the polymer is recovered rapidly without the thermal degradation which can occur with flash devolatilization. In addition, judicious selection of the non-solvent and mixing conditions can result in a precipitate which is easily filtered and dried.

PET Processing Train. In one example of a combined technology process, post-consumer 2–liter bottle material is divided into polyolefin and poly(ethylene terephthalate) fractions using water in sink-float tanks followed by treatment of these fractions in solvent-processing trains. Research on the interactions of polyolefins with xylene (20-26) indicates that this solvent would be suitable for use in the polyolefin solvent treatment train. Therefore, the selection and evaluation of solvents for the PET processing train has been the major focus of the present research (27). The selection of a process solvent for the PET train was based on the criteria of: PET-solvent compatibility, HDPE-solvent incompatibility, toxicity, cost, and ability to recover solvent for reuse in the process. The solvent of choice for PET is N-methyl-2-pyrrolidinone (NMP). NMP is of low toxicity (28-30), is biodegradable (aerobic wastewater treatment), is easily recovered, and is commonly used in many processes (31). In addition, NMP does not appreciably dissolve HDPE, even at elevated temperatures (13, 27) although it does readily dissolve PET at these same temperatures.

Role of the Crystalline Nature of PET. In order to determine the optimum operating conditions for the PET processing train, the dissolution and crystallization behavior of PET in NMP must be analyzed. To this end, the dissolution rate of 2-liter bottle PET in NMP as a function of solvent temperature was determined from the mass loss of sections of 2-liter bottles exposed to NMP at various temperatures for a range of exposure times (27). As shown in Figure 3, NMP was found to rapidly dissolve PET sections of 2-liter bottles at temperatures greater than 130°C. In fact, at 160°C, the sections dissolved in approximately 5 minutes. Below 130°C, it was found that the sections swelled, but did not dissolve. This dramatic difference in dissolution behavior over a small temperature range is due to the presence of crystallites in the PET which have relatively distinct dissolution temperatures. The crystallites act as cross–links below the crystal dissolution temperature (T_d), allowing the polymer to swell, but not to dissolve, despite the fact that the amorphous regions would prefer to be dispersed in the solvent. Once the PET sections are heated above T_d, the crystallites dissolve and the polymer readily disperses into solution.

This dissolution rate information implies that in the first stage of the PET processing train (Solvent Washing), NMP at temperatures as high as 130°C can be used to remove soluble impurities (adhesives, coatings, and PS, PVC, and polycarbonate (PC) contamination), while leaving the PET sections only slightly swollen. In the second stage (Target Polymer Dissolution), the PET chips can be mixed with recycled NMP at 160°C, bringing about the rapid dissolution of the sections without appreciably dissolving any HDPE which may have left the sink-float process in the the PET stream.

The main impurities in the PET processing train will most likely be aluminum and HDPE. As mentioned previously, aluminum removal is easily achieved by filtering the low-viscosity polymer solution. This alleviates the need for low-efficiency melt filtration units or expensive eddy current separators which represent a large fraction of plant costs for non-solvent separation processes (*32*). Ideally, HDPE should not leave the sink-float unit in the PET stream. However, because the sink-float system is not 100% effective, small amounts of HDPE (and some PP) will be present. The advantage of using NMP in the PET processing train is that HDPE is only slightly soluble at elevated temperatures as indicated by the data in Figure 4. For example, at 160°C, only about 0.1 wt% of the HDPE will dissolve in 30 minutes. At this same temperature, PET sections of 2-liter bottles would be completely dissolved in about 5 minutes. The presence of PVC in recycled PET is also a concern because of the damage caused to injection molding equipment exposed to PVC at PET processing temperatures as well as color formation in the plastic. In a selective dissolution process using NMP, trace amounts of PVC should be removed in the Solvent Washing stage since PVC is quite soluble in NMP at 120°C.

The later stages of the PET processing train (Solution Purification and Polymer Recovery) are also impacted by the crystalline nature of PET. If cooled below T_d, PET will crystallize from solution at a rate which is controlled by nucleation and growth kinetics. In the purification stage, it may be necessary to cool the solution before exposing the separation medium to hot NMP, but crystallization of the polymer is not desired. Also, if temperature quenching is used to recover PET from solution, the rate of recovery is controlled by the crystallization kinetics. In addition, even if a non-solvent is added to recover the PET, the solution may need to be cooled prior to the addition of the non-solvent. It is necessary that crystallization be avoided in this cooling process to ensure that the heat exchanger does not become plugged with precipitate. Therefore, it is critical that the quench crystallization kinetics of PET from solution in NMP be studied.

In this paper, the temperature, concentration, and time dependence of crystallization of a commercial PET sample from solution in NMP will be described using data from dilatometric and turbidimetric techniques combined with established crystallization rate relationships.

Experimental

All of the samples described were prepared in an inert atmosphere (nitrogen or argon) with vacuum dried Goodyear Cleartuf 7207 PET resin [PET(7207)] and Aldrich anhydrous NMP. The turbidimeter samples were flame-sealed in 28 mm OD Pyrex cells which held about 35 mL of solution. Flame-sealed samples with concentrations ranging from 0.488 to 9.79 weight % PET(7207) in NMP were studied.

A dilatometer was used to monitor the overall extent-of-crystallization and was designed to hold ≈80 mL of solution. The sample chamber was constructed of 38 mm OD pyrex tubing. The details of the dilatometer used in this study have been described elsewhere (*27*). Three dilatometer samples were analyzed with concentrations of 0.78, 1.7, and 3.3 weight % PET(7207) in NMP.

Turbidimetric Crystallization Time and Dissolution Temperature Measurements. When a polymer precipitates or crystallizes from solution, the

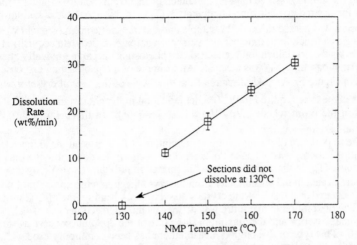

Figure 3. Effect of solvent temperature on the dissolution rate of PET sections from 2-liter bottles. Error bars represent range of measured rates. (Adapted from ref. 42)

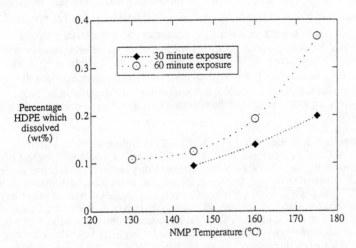

Figure 4. Effect of solvent temperature and exposure time on the extent of dissolution of HDPE in NMP. (Adapted from ref. 42)

originally transparent medium becomes cloudy (turbid). Therefore, the early stages of polymer crystallization from solution can be monitored by using a modified cloud point apparatus (turbidimeter). To initiate the experiment, a sample was heated in a 170°C silicone oil bath to dissolve the PET, and then quenched in the turbidimeter which was held at the desired crystallization temperature (T_c). In this apparatus, a helium-neon laser beam was passed through the sample and the transmitted photosignal (V) was recorded as a function of time using a personal-computer-based data acquisition system. As the crystallization proceeded, the photosignal became gradually attenuated until a final baseline (V_f) was reached. An example of the photosignal response is shown in Figure 5. A good measure of the time scale of the overall crystallization process is the time required for a 20% net reduction in the photosignal ($t_{0.2}^{ph}$). This occurs at the point where ($V_i - V$) = 0.2($V_i - V_f$). In this relationship, V_i is the initial baseline voltage.

The importance of the dissolution temperature of PET crystallites to the selective dissolution process was discussed in a previous section. To determine the dissolution temperature (T_d) of the crystalline material formed in turbidimetric experiments, the cells were heated in the turbidimeter from the crystallization temperature until the solution cleared as a result of the crystals dissolving. A heating rate of 1°C/min was used. The dissolution temperature was defined as the intersection of the final photosignal baseline and the maximum slope line of the dissolution curve as shown in Figure 6. This type of dissolution measurement has been commonly used (26,33,34). This dissolution temperature is generally a linear function of the crystallization temperature (T_c). Another parameter, the equilibrium dissolution temperature (T_d^o), is defined as the intersection of the T_d vs. T_c line with the $T_d = T_c$ line as shown in Figure 7 for several of the flame-sealed samples. Theoretically, at this temperature, T_d^o, the solution is in crystallization/dissolution equilibrium and the crystallization of PET will take an infinite amount of time. It is the distance from T_d^o ($\Delta T = T_d^o - T_c$) which controls the rate of crystallization. For a given polymer concentration, a slower crystallization rate yields a more perfect and stable crystalline structure which will have a higher dissolution temperature. Therefore, T_d should rise as T_c is raised. In addition, if the polymer concentration is increased, the solution thermodynamics are altered (chemical potential of the crystals and solution are changed), leading to an increase in the dissolution temperature as evidenced by the data in Figure 7.

Crystallization Kinetics Measurements - Dilatometry. In this method, the development of crystallinity in a PET(7207)/NMP solution was monitored utilizing dilatometric techniques. The theory behind dilatometry is that crystallization leads to an increase in sample density and, therefore, the volume of the sample decreases. The dilatometer contains a confining liquid (often mercury) which is open to the atmosphere via a graduated capillary. As the sample volume decreases, so does the height of the confining liquid in the capillary. Therefore, the extent-of-transformation at any time, X(t), (crystallization in this case) can be related to the capillary liquid height change using the following expression:

$$X(t) = \frac{(h_o - h_t)}{(h_o - h_\infty)} \tag{1}$$

where h_o is the initial mercury height, h_t is the height at time t, and h_∞ is the final height.

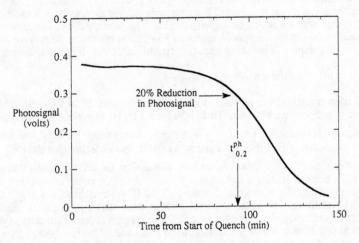

Figure 5. Response of the transmitted photosignal to the crystallization at 95°C of a 3.97 wt% PET(7207) in NMP flame sealed sample.

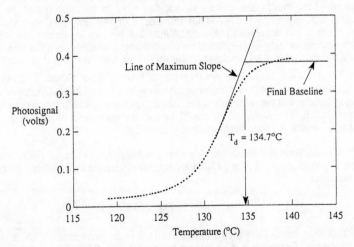

Figure 6. Determination of the dissolution temperature (T_d) for the 3.97 wt% PET(7207) in NMP sample crystallized at 95°C. Heating rate: 1°C/min.

As was the case for the flame-sealed samples, the dilatometer was heated to 170°C in a large silicone oil bath to dissolve the PET followed by a rapid quench in a second bath to the desired crystallization temperature. As crystallization proceeded, the height of mercury (h_t) in the graduated capillary column of the dilatometer was recorded The extent-of-transformation (X) was calculated from equation 1. In addition, for these dilatometer samples, the transmitted laser photosignal was monitored in a manner similar to that used for the flame-sealed samples. The primary difference between the two was that the dilatometer path length was 35 mm as opposed to 26 mm for the flame-sealed samples. This should not significantly alter the photosignal response.

Results and Discussion

The solution crystallization process, much like crystallization from the melt, is highly sensitive to temperature changes. This is evidenced by the exponential variation of the 20% photosignal reduction time ($t_{0.2}^{ph}$) with quench temperature for the flame-sealed samples as shown in Figure 8. It is also apparent from this figure that the PET concentration affects $t_{0.2}^{ph}$ almost as dramatically as does the crystallization temperature. In addition, because it is a nucleation and growth phenomenon, the crystallinity develops exponentially with time. For the PET/NMP system, these two properties result in crystallinity development curves such as those shown in Figure 9. Each curve in Figure 9 is composed of three regions. The first region is one of primary crystallite nucleation and growth which is too small to be measured. This is followed by a region of auto-catalytic growth in which the growing surface area of the crystals causes the crystallization to accelerate. Finally, the primary crystallization rate declines, often followed by a slower secondary crystallization process.

The onset of the auto-catalytic region is not easily determined from dilatometric data because of the small volume change associated with the initial stages of the process. Fortunately, because even small amounts of crystallinity produces a cloudy suspension, the initial development of crystallinity can be detected turbidimetrically, as was discussed previously. In Figure 10, a comparison of the photosignal and dilatometer responses to developing crystallinity for the same sample, it is seen that at $t_{0.2}^{ph} = 670$ minutes the dilatometer height registers little, if any, change. It is evident that the transmitted photosignal is more sensitive to the onset of crystallization than is the mercury height, but the dilatometer continues to provide crystallization data after the photosignal is fully attenuated. As a result, the two methods yield information on two different time scales and are, therefore, complimentary.

Solution Crystallization Kinetics. The crystallization kinetics of polymers from the melt (35-38) or from solution (23,35) are commonly interpreted in terms of the Avrami equation:

$$X(t) = 1 - \exp[-kt^n]$$ (2)

where X(t) is the extent-of-crystallization at time t, k is the rate constant, and n is the Avrami exponent. In theory, n is an integer between 1 and 4 depending on the mode of nucleation and crystal growth. Three-dimensional growth resulting from homogeneous nucleation exhibits n = 4 behavior while two dimensional growth results in n = 3. Non-integer values for n may result from heterogeneous nucleation. The Avrami expression is generally applicable for the initial stages of crystallization, but as impingement of growing crystals occurs, results deviate from the basic Avrami equation. In dilute and semi-dilute polymer solutions, the impingement effect of

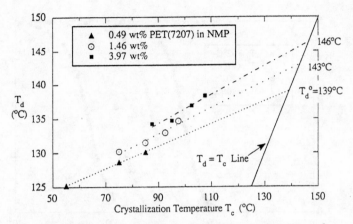

Figure 7. Determination of the equilibrium dissolution temperature (T_d^0) from a plot of dissolution temperature (T_d) vs. crystallization temperature (T_c) for three flame sealed samples of PET(7207) in NMP.

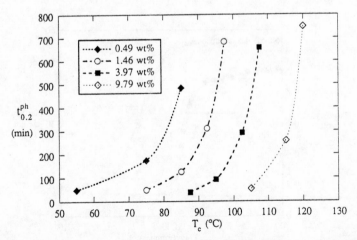

Figure 8. Effect of crystallization temperature (T_c) and polymer concentration on the 20% photosignal reduction time ($t_{0.2}^{ph}$) for flame sealed samples of PET(7207) in NMP.

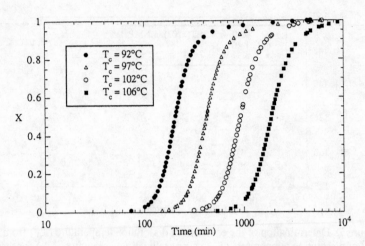

Figure 9. Plot of fraction crystallized (X) as a function of time and quench temperature (T_c) for 3.3 wt% PET(7207) in NMP dilatometer sample.

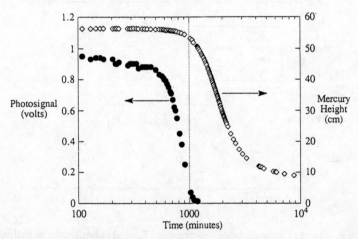

Figure 10. Variation of the mercury height and transmitted photosignal with time for a dilatometric solution crystallization experiment.
Sample: 3.3 wt% PET(7207) in NMP crystallized at 106.0°C.

growing centers is lessened due to the lower concentration of crystalline material. As a result, the Avrami equation generally applies over a wider extent-of-transformation range for polymer solutions than for pure polymers (23,35).

Over the years, many modified Avrami relationships have been proposed to account for secondary crystallization (39,40) and parallel crystallization processes (41). In this work, it was found that secondary crystallization did account for a measurable fraction of the final crystallinity. This being the case, equation 2 must be modified to account for the contributions of both the primary and the secondary processes:

$$X_{total} = \omega_1 X_1 + (1 - \omega_1)X_2 \tag{3}$$

where ω_1 is the fraction of the final crystallinity due to the primary process and X_i is the extent-of-crystallization $(0 \leq X_i \leq 1)$ for the ith process. For short times $(X_{total} < 0.6)$, the primary process dominates the overall crystallization and can be described using equation 2. Therefore, for less than 60% total transformation:

$$X_{total} = \frac{(h_0 - h_t)}{(h_0 - h_\infty)} \approx \omega_1 X_1 = \omega_1(1 - e^{-kt^n}) \tag{4}$$

When the data for the three dilatometer samples were fitted to equation 4, it was determined that the exponent $n = 4$ described the primary crystallization most effectively and that approximately 75% of the total crystallinity was a result of the primary process $(\omega_1 \approx 0.75)$.

The rate constant (k), weighting factor (ω_1) and Avrami exponent (n) in equation 4 can be determined from the $X < 0.6$ dilatometer data using a curve fitting procedure. Unfortunately, this method requires knowledge of the final dilatometer height (h_∞) which can take weeks to achieve even when the majority of the crystallization occurs in the first 24 hours. To alleviate the need for h_∞, a novel derivative method was devised. If the time derivative of equation 4 is divided by X_{total} the following expression results:

$$\frac{1}{X_{total}} \frac{dX_{total}}{dt} = \frac{-dh/dt}{(h_0 - h_t)} = \frac{nkt^{n-1}}{e^{kt^n} - 1} \tag{5}$$

Not only has the final height been removed from the relationship, but the weighting factor, ω_1, is also no longer present. Therefore, only data from the first stage of the process are necessary to determine k and n. Also, by using equation 5, the number of fitted constants has been reduced from three to two. In the present study, using an Avrami exponent $n = 4$, the values for k calculated from equation 4 and equation 5 differed by less than 10%. The average values of the rate constant are presented in Table I along with values of $t_{0.2}^{ph}$ determined for the dilatometer samples. It is clear from the data in the table that both k and $t_{0.2}^{ph}$ are highly sensitive to temperature and concentration.

As shown above, the dilatometer is a versatile tool for characterizing the crystallization kinetics of polymer solutions. Unfortunately, operation of the dilatometer requires constant attention for extended periods of time, during which, precise temperature control must be maintained. These restrictions limit the number of

Table I: Crystallization data for dilatometer samples

Concentration (wt% PET)	Crystallization Temperature (°C)	20% Photosignal Reduction Time (min)	Avrami Rate Constant (k) (min^{-4})
0.78	65.0	42	1.5e-08
"	75.0	87	4.8e-10
"	82.5	190	3.3e-11
"	89.0	400	1.6e-12
1.7	77.5	45	4.1e-09
"	84.9	98	2.2e-10
"	92.5	260	4.7e-12
"	97.5	520	3.0e-13
3.3	92.0	80	4.8e-10
"	97.0	160	2.9e-11
"	102.0	350	1.4e-12
"	106.0	670	9.6e-14

samples which can be studied. Conversely, turbidimetric crystallization experiments do not require strict temperature control and can be monitored with data acquisition systems. In addition, because the reduction of the photosignal occurs in a much shorter time frame than the overall crystallization process, many more samples can be analyzed. The turbidimeter apparatus can also be used to determine the dissolution temperatures on the same samples. For these reasons, photosignal crystallization data was correlated to dilatometer extent-of-crystallization data. Using data from these two methods, it was found that the ratio of the time required for 10% transformation in the primary process $(X_1 = 0.1)$, $t_{0.1}$, to the 20% photosignal reduction time, $t_{0.2}^{ph}$, was constant for a given sample, independent of the crystallization temperature. Therefore the following relation can be written:

$$t_{0.1} = \Phi \, t_{0.2}^{ph} \tag{6}$$

where Φ is a constant with respect to temperature, but is somewhat concentration dependent. This indicates that $t_{0.2}^{ph}$ is indeed a good measure of the time scale of the solution crystallization process.

Even more dramatic is the fact that $t_{0.2}^{ph}$ can be directly related to the the rate constant k as shown in Figure 11. According to Figure 11 (and predicted by combining equation 2 and equation 6), k is related to $t_{0.2}^{ph}$ via the following relationship:

$$k = \left(\frac{-\ln(0.9)}{\Phi^n} \right) \left(\frac{1}{t_{0.2}^{ph}} \right)^n \tag{7}$$

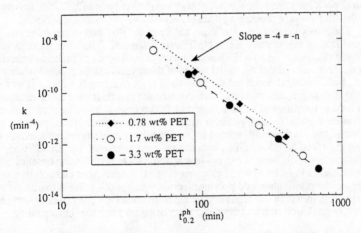

Figure 11. Power-law relationship between the Avrami rate constant (k) and the 20% photosignal reduction time ($t_{0.2}^{ph}$) for the PET(7207)/NMP dilatometer samples. Curve fit: $k = A\,(t_{0.2}^{ph})^{-4}$ (A = constant with respect to temperature).

where n is the Avrami exponent. This is an extremely powerful relationship because only a single set of dilatometer height response data is required for a given sample in order to determine Φ. Afterwards, the rate constant can be directly calculated using data from the much simpler turbidimetric experiments.

Conclusions

The use of a combined technology process to separate mixed plastics has been described. It is proposed that a sink float process be used as a "first cut" yielding multiple streams of plastics. These streams can then be further purified with solvent processing trains in which the process solvents have been optimized for the removal o one target polymer or a specific family of polymers. The strength of a sink-float system is cost-effectiveness, but the system is limited by the relatively low purity of th final products. On the other hand, selective dissolution processes are capable of producing polymers of high purity, but at an increased cost. 2–liter bottles make an attractive example because of their post-consumer availability (primarily from states with deposit laws) and the relatively high value of recycled PET. The solvent of choic for the PET solvent processing train is N-methyl-2-pyrrolidinone (NMP).

The importance of the crystalline nature of the PET and of the solution crystallization kinetics of PET in NMP were discussed. The latter was characterized utilizing dilatometric and turbidimetric techniques. The turbidimetric method provided more information about the initial stages of the crystallization process while the dilatometric method described the long-term crystallization behavior. It was shown th the crystallization data measured turbidimetrically can be related to the dilatometer data and used to determine the rate constant of the process. The simpler turbidimetric technique allowed for analysis of a wider range of samples. It is apparent that a turbidimetric device could be used to pre-test the crystallization characteristics of a given PET feed and determine the optimum process conditions. In addition, turbidimeters could serve as an on–line analysis tools to detect the onset of crystallization in order to avoid clogging heat exchanger tubes and solution purificatior filters in the purification and polymer recovery stages of a solvent treatment train, or any process in which materials crystallize from solution. The effect of solvent and PE structure on dissolution and crystallization behavior have also been studied.

Acknowledgements

The authors wish to thank the Plastics Institute of America for research fellowships, th National Science Foundation and DuPont for stipend support, and the Goodyear Tire (Rubber Co. for PET samples.

Literature Cited

1. Sperber, R.J.; Rosen,S.L. *SPE ANTEC Tech. Papers*, **1975**, *21*, 521.
2. Sperber, R.J.; Rosen,S.L. *Polym. Eng. Sci.*, **1976**, *16*, 246.
3. Seymour, R.B.; Stahl, G.A. *J. Chem. Ed.*, **1976**, *53*, 653.
4. Murphy, W.R.; Otterburn, M.S.; Ward, J.A. *Polymer*, **1979**, *20*, 333.
5. Meyer, M.F., Jr.; Combs, R.L.; Wooten, W.C., Jr.; (to Eastman Kodak), US Pat. 3,701,741 (Oct 31, 1972).
6. Kajimoto, H.; Shimada, T.; Tokuda, M.; Tamura, T.; (to Mitsubishi Heavy Industries) Japan. Kokai Pat. 76 16,378 (Feb 9,1976).
7. Sidebotham, N.C.; Young, C.W.; Shoemaker, P.D.; (to Monsanto), US Pat. 4,003,881 (Jan 18, 1977).
8. Nishimoto, Y.; Mizumoto, Y.; Shigeo, H.; Mitsuoka, S.; (to Mitsubishi Heavy Industries), Japan. Kokai Pat. 76 20,976 (Feb 19,1976).

9. Sidebotham, N.C., Young, C.W.; Shoemaker, P.D.; (to Monsanto), US Pat. 4,137,393 (Jan 30, 1979).
10. Polaczek, J.; Pielichowski, J.; Zygmunt, L. *Fuel,* **1987**, *66,* 1556.
11. Kampouris, E.M.; Papaspyrides, C.D.; Lekakou, C.N. *Polym. Eng. Sci.,* **1988**, *28,* 534.
12. Lynch,J.C.; Nauman, E.B.; presented at *SPE RETEC: New Developments in Plastic Recycling,* Oct. 30-31, 1989.
13. Vane, L.M.; Rodriguez, F.; presented at *SPE RETEC: Recycling of Plastic Materials to Meet Government and Industry Requirements,* November 29-30, 1990.
14. Vane, L.M.; Rodriguez, F.; presented at *ACS Polymer Technology Conference,* June 2-5, 1991.
15. *Mod. Plastics,* August **1990**, 126.
16. *Mod. Plastics,* November **1988**, 22.
17. *Plastics Eng.,* March **1990**, 81.
18. *Chem. & Eng. News,* January 30, **1989**, 7.
19. Plastics Recycling Foundation brochure: *"Plastics Recycling: A Strategic Vision".*
20. Sittig, M.; *Organic and Polymer Waste Reclaiming Encyclopedia;* Noyes Data Corp.: Park Ridge, NJ, 1981; 327 and 357.
21. Prasad, A.; Mandelkern, L.; *Macromolecules,***1989**, *22,* 914.
22. Mandelkern, L.; *Polymer Preprints,* **1986**, *27,* 206.
23. Devoy, C.; Mandelkern, L.; Bourland, L.; *J. Polym. Sci. A-2,* **1970**, *8,* 869.
24. Mandelkern, L.; *Polymer,* **1964**, *5,* 637.
25. Drain, K.F.; Murphy, W.R.; Otterburn, M.S.; *Conservation and Recycling,* **1981**, *4,* 201.
26. Jackson, J.F.; Mandelkern, L.; *Macromolecules,* **1968**, *1,* 546.
27. Vane, L.M.; Ph.D. Thesis, Cornell University, Ithaca, NY, 1992.
28. Lee, K.P.; Chromey, N.C.; Culik, R.; Barnes, J.R.; Schneider, P.W.; *Fundam. Appl. Toxicol.,* **1987**, *9,* 222.
29. *Rapid Guide to Hazardous Chemicals in the Workplace;* Sax, N.I.; Lewis, R.J., Sr., Eds.; Van Nostrand Reinhold Co.: NY, NY, 1986.
30. Wells, D.A.; Thomas, H.F.; Digenis, G.A.; *J. Appl. Toxicol.,* **1988**, *8,* 135.
31. *M-Pyrol N-Methyl-2-Pyrrolidone Handbook;* GAF Corp., 1972.
32. Thayer, A.M.; *C & E News,* Jan. 30, 1989, p. 7.
33. Lemstra, P.J.; Challa, G. *J. Polym. Sci.: Polym. Phys. Ed.,* **1975**, *13,* 1809.
34. Minagawa, M. et al. *Macromolecules,* **1989**, *22,* 2054.
35. Mandelkern, L. *Crystallization of Polymers,* McGraw-Hill, NY,NY, 1964.
36. Ciora, R.J., Jr.; Magill, J.H. *Macromolecules,* **1990**, *23,* 2350.
37. Avrami,M. *J. Chem. Phys.,* **1939**, 7, 1103; ibid. **1940**, *8,* 1212; ibid. **1941**, *9,* 177.
38. Sharples, A. *Introduction to Polymer Crystallization,* Edward Arnold, London (1966).
39. Hay, J.N; Mills, P.J. *Polymer,* **1982**, *23,* 1380.
40. A. Booth and J.N. Hay, *Polymer,* **1971**, *12,* 365.
41. P. Cebe, *Polym Eng. & Sci.,* **1988**, *28,* 1192.
42. Rodriguez, F.; Vane, L.M.; Schlueter, J.J.; Clark, P.; in *Separation Science in Environmental Chemistry - ACS Symposium Series Volume,* American Chemical Society (in preparation).

RECEIVED April 29, 1992

Chapter 12

Evaluation of Polymer Degradation in Controlled Microbial Chemostats

Gary L. Loomis[1], James A. Romesser[2], and William J. Jewell[3]

[1]Warner-Lambert Company, 201 Tabor Road, Morris Plains, NJ 07950
[2]Celgene Corporation, 7 Powderhorn Drive, Warren, NJ 07060
[3]Microgen Corporation, Ithaca, NY 14853

We have developed protocols for the assessment of polymer degradation utilizing stable, well-controlled microbial chemostats (digesters) with easily measurable and quantifiable populations of microorganisms. This paper describes the use of three different chemostats [thermophilic (60°C) anaerobic, thermophilic (60°C) aerobic, and mesophilic (35°C) aerobic] to assess the degradation of a variety of commercially available polymer films.

Environmental concerns have recently triggered a flurry of new research into the environmental fate of polymers.[1,2,3,4] Considering the importance of the subject, surprisingly little is understood about the molecular-level interactions of many common polymers with microorganisms. The varying claims regarding the degradation of these materials may be due in part to inconsistencies in the use of terminology describing degradation and to the lack of accepted standard tests for assessing the biodegradability of plastics.[5] Most of the existing methodology for the study of polymer degradation suffers in the utilization of uncontrolled and poorly defined biological milieu -- e.g. soil[6], compost[7], sewage sludge[8], sea water, etc. can have highly variable compositions in terms of both chemistry and microbial populations. Commonly used evaluative methods based solely on physical properties (i.e. tensile strength, % elongation, etc.) may often be unsuitable, since observed property changes can be effected by changes in polymer morphology and, therefore, are not necessarily a measure of degradation. Conclusions about polymer degradation based solely on respiration and biogas production data is often inconsistent[9,10], possibly because mechanisms other than digestion of the polymer component of a system can alter the respiration of microorganisms. Finally, assessment of material biodegradability by observation of the degree of microbial growth[11] has been often inconclusive, since common polymer additives may serve as a nutrient source while the molecular structure of the polymer remains intact.

0097–6156/92/0513–0163$06.00/0
© 1992 American Chemical Society

EXPERIMENTAL:

The standard feed for all three reactors was composed of cellulose and sorghum[12], with the sorghum being the main source of nutrients, and a supplemental solution of trace nutrients. Reactors were semi-continuously fed twice per week and a relatively low organic loading rate (OLR) of 1.42 g of VS per 1.0 kg of reactor mass per day was maintained in order to insure equal exposure of the sample materials to the microbes while providing a significant microbial population. Volatile solids (VS) is a standard, surrogate, operational measure of the organic fraction of a complex organic material or mixture and is based on the fact that most organic compounds volatilize at 550°C in air[13]. The chemostats were run in a slurry mode maintained at 4.5 wt.% dry solids at near neutral pH. The biological activity of these system was monitored via analysis of quantity and composition of biogas and volatile fatty acids (VFA) produced for the anaerobic system and CO_2 evolution for the aerobic systems.

Polymer samples were in the form of 2.0 cm X 10.0 cm strips of film. These film samples had a thickness of between 0.5 and 3.0 mils with most materials being in the range of 1.5 to 2.0 mils. Films were immersed in the reactors for up to 28 days and retrieved materials were evaluated by a variety of analytical techniques with particular attention paid to changes in molecular weight and molecular weight distribution. Obviously, since the microorganisms in this digester were fed with sorghum as the major carbon source and the polymer samples represent only a minor portion of the total available carbon, it is not possible to accurately assess the degradation of the samples by monitoring biogas (methane in the anaerobic chemostat and carbon dioxide in the aerobic chemostats) production. Simultaneously, two sets of control samples were maintained; one set in sterile, phosphate buffer at pH=7.5 at the appropriate temperature, and a second set in a dry environment at 20°C.

Molecular weight data was obtained using Waters model 150°C gel permeation chromatographs (GPC). Polyolefin materials were run in 1,2,4-trichlorobenzene at 135°C using two Shodex AT-80M/S columns with calibration as polyethylene via the universal calibration technique using polystyrene as primary standard. Polyester materials were run at 35°C with a mobile phase of 0.01 M sodium trifluoroacetate in hexafluoroisopropanol and two Shodex KF80M/HFIP columns with an in-house laboratory standard of poly(ethylene terephthalate) used for calibration.

The stress-strain data presented in the accompanying bar graphs were generated from Instron testing using the average of five specimens of each sample, and the error bars (only the top half of the bar is shown) represent the standard error as $(S/N^{1/2})$.

DISCUSSION AND RESULTS:

Cellophane:

Uncoated cellophane (regenerated cellulose) film was used as a positive control material and since, due to the high cellulose content of the feed, the digesters are rich in microorganisms which secrete cellulases (cellulose degrading enzymes), it was expected that the cellophane would degrade readily within the 28 day duration of the experiments. This was indeed the case, with no trace of the cellophane recoverable from any of the three reactors. Commercial cellophanes, which are generally coated with a thin, moisture-barrier coating of either nitrocellulose or poly(vinylidene chloride) were not evaluated in this study.

Polyolefins:

Figures 1 and 2 show stress-strain data for films of high density polyethylene (Alathon A7030), linear low density polyethylene (Dowlex 2045), two polyolefins before and after exposure to the anaerobic system. These property changes, particularly the HDPE, (given the inconsistencies inherent in testing physical properties of films) are significantly beyond experimental error and are therefore probably real. However, since the GPC traces shown in Figures 3 and 4 indicate no change in molecular weight distribution, we conclude that no significant molecular level degradation (i.e. no chain scission) has occurred. The physical property changes are likely due to changes in orientation or crystal morphology of the polymer caused by the 28 day 60°C "thermal exposure". Various commercial "plastic" bags were evaluated with similar results.

Polyesters:

Figures 1 and 2 (stress-strain data) and Figures 5 and 6 (molecular weight data) show a comparison between the polyesters PET (Mylar™, from DuPont) and poly (hydroxybutyrate/valerate) copolymer (Biopol™ from ICI). It can be easily seen that, in the anaerobic system, a film of the commercial Biopol™ shows no significant physical property change while showing substantial loss of molecular weight. Notice that the Mylar™, on the other hand, shows significant changes in physical properties under the same conditions with no accompanying change in molecular weight (therefore no molecular level degradation). Like the polyethylene examples above, the changes in physical properties of the Mylar™ are likely due to changes in morphology, in this case possibly loss of orientation. The apparent degradation of the Biopol™ without measurable changes in stress-strain properties is likely due to the high initial molecular weight (M_W >400K) of this microbially produced polyester and further exposure would no doubt cause the material to exhibit property deterioration. It must be pointed out that we have no direct evidence that the molecular weight loss of the poly(hydroxy-butyrate/valerate) in our systems is due to the action of microorganisms -- we may well be seeing simple hydrolysis. It is also interesting, but not yet explainable, that the hydrolysis of the Biopol™ film appears to be faster in pH 7.5 phosphate buffer than in the anaerobic chemostat (see Figure 6).

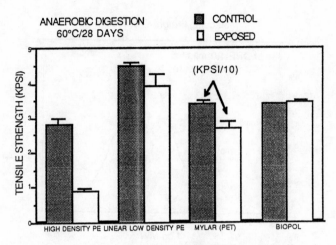

Figure 1. Tensile strength before and after anaerobic digestion.

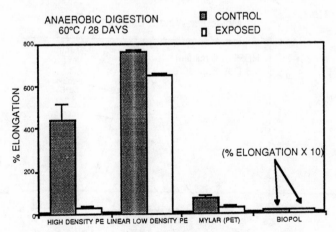

Figure 2. Elongation before and after anaerobic digestion.

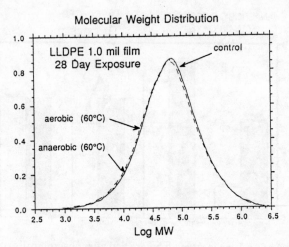

Figure 3. Molecular weight distribution of LLDPE before and after digestion.

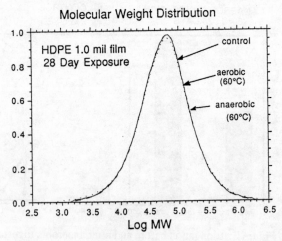

Figure 4. Molecular weight distribution of HDPE before and after digestion.

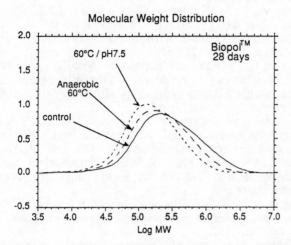

Figure 5. Molecular weight distribution of Biopol™ before and after digestion.

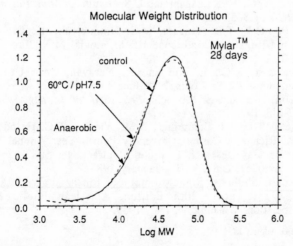

Figure 6. Molecular weight distribution of Mylar™ before and after digestion.

References

1 Doi, Y., Y. Kanesawa and M. Kunioka (1990), *Macromolecules,* Vol. 23, pp 26-31,1990.
2 Hosokawa, J., M. Nishiyama, K. Yoshihara and T. Kubo , *Ind. Eng. Chem. Res.,*Vol.29, No. 5, pp 800-805, 1990.
3 Albertsson, A-C. and S. Karlsson (1990), *Prog. Polym. Sci.,* Vol. 15, No. 2, pp 177-192, 1990.
4 Iannotti, G., N. Fair, M. Tempesta, H. Neibling, F.H. Hsieh and R. Mueller paper presented at the First International Degradable Plastics Workshop, Toronto, Ontario, Canada, November 1-4, 7 pages, 1989.
5 The American Society for Testing and Materials is currently proposing the use of both anaerobic and aerobic bioreactors as standard test environments: ASTM 1990a and ASTM 1990b are proposed test protocols under development by ASTM Subcommittee D-20.96 on Environ-mentally Degradable Plastics, 1990.
6 Ong, K.C. and W.R. Stanton, *Sago-76: Papers of the First International Sago Symposium* Ed.) Koonlin Tan, Kuala Lumpur, pp 240-243, 1987.
7 Maddever, W.J. and G.M. Chapman, *Proceedings of Symposium on Degradable Plastics,* The Society of the Plastics Industry, Inc., Washington, DC, June 10, 1987.
8 Kumar, G.S., V. Kalpagam and U.S. Nandi. *J. Appl. Polym. Sci.,* Vol 26, pp 3633-3641, 1981.
9 Rivard, C.J., T.B. Vinzant, M.E. Himmel and K. Grohmann, Corn Utilization Conference III, St. Louis, MO, June 19-21, 1990.
10 Srivastava, K.C., L. Nie and R. Narayan, Corn Utilization Conference III, St. Louis, MO, June 19-21, 1990.
11 Griffin, G.L. and H. Mivetchi, *Proc. 3rd Int. Biodeg. Symp.,* pp 807-813, 1976.
12 Jewell, W.J., R.J. Cummings, A.M. Whitney, F.G. Herndon and B.K. Richards, Cornell University Fourth Year Annual Report for the Gas Research Institute, Contract No. 5083-226-0848, GRI-87/0061, Chicago, IL, January 1987.
13 Standard Methods, Standard methods for the analysis of water and wastewater, 20th Edition, American Public Health Association, Washington, DC., 1989.

RECEIVED August 19, 1992

POLYMER RECOVERY

Chapter 13

Separation of Thermoplastics by Density Using Near-Critical and Supercritical Carbon Dioxide and Sulfur Hexafluoride

M. S. Super, R. M. Enick, and E. J. Beckman

Chemical Engineering Department, University of Pittsburgh, Pittsburgh, PA 15261

Near-critical and supercritical fluids composed of carbon dioxide and/or sulfur hexafluoride have been used to sort thermoplastic waste mixtures according to density. For example, PVC can be readily removed from waste PET, and tinted or filled materials can be separated from their clear and unfilled counterparts. Carbon dioxide alone can be used to separate polyolefins, while pure sulfur hexafluoride, an extremely dense fluid, can be used to separate the non-olefin thermoplastics. Sulfur hexafluoride-rich CO_2/SF_6 mixtures can be used to separate all of the thermoplastics. The brief exposure of the thermoplastics to the mild temperature, high-pressure environment did not chemically or physically alter them. The densities of the supercritical or near-critical fluid mixtures have been accurately correlated to temperature, pressure and fluid composition using a cubic equation of state.

Strategies for secondary recycling of plastics can be divided into two major areas: (1) those emphasizing material separation and ultimate fabrication of similar end-use products; and (2) those devoted to reprocessing of co-mingled waste to replace concrete and wood in products which do not require the physical properties of the virgin materials {1-4}. Naturally, co-mingled polymeric waste is often contaminated with other materials, such as wood, paper, metals (both ferrous and nonferrous) and glass. Removal of these non-polymeric components is necessary both to protect equipment from harmful abrasives and to achieve physical properties in the reprocessed material which are reasonably close to those of virgin resin. In general, experience has shown that physical properties, and thus resale value, of recycled polymeric materials

0097–6156/92/0513–0172$06.00/0
© 1992 American Chemical Society

increases as the purity of the material increases. Furthermore, certain thermoplastic mixtures, such as polyvinylchloride ("PVC") and polyethylene terephthalate ("PET"), or of crosslinked and linear polyethylene, can lead to material and/or equipment degradation when reprocessed together. Within the same general class of polymer, separation by color will also generally raise the value of the recycled product. Although not deleterious to physical properties, re-processing of mixed dark and light-colored polyolefins can limit their re-use in layered virgin/recycled "composites", while un-tinted PET commands a significantly higher re-sale price than a mixture of green and clear material. Consequently, schemes capable of selectively separating each polymeric component from a co-mingled mixture enhance the value of waste thermoplastics.

Synthetic polymer waste streams are composed primarily of high and low density polyethylene ("HDPE" and "LDPE"), polypropylene ("PP"), polystyrene ("PS"), both in foamed and bulk form, PET, and PVC. The recycling of these thermoplastics has been limited by difficulties in separating the polymers from each other and from non-polymeric contamination. Current schemes to separate thermoplastic waste generally rely on hand-sorting or mechanical sortation for whole bottles, and either hydrocloning or air classification for granulated waste. Several commercially vital separations of granulated thermoplastic waste, such as PVC from PET, or classification of the olefin component of the waste stream, are not efficiently performed using current technology. Furthermore, component selectivity in both air classifiers and hydrocyclones is a function of particle size distribution, as well as particle density, which limits overall separation efficiency {5}.

Proposed Separation Method

In this paper a novel process is outlined for the density-based separation of not only waste thermoplastics from metal, glass, paper and wood contaminants, but also the individual polymers from each other. This new process employs the unique properties of a fluid near its critical point to allow fine separations at mild temperatures and pressures. These separations are achieved simply by varying the density of a supercritical or near-critical fluid to a value which is intermediate to the densities of the thermoplastics. The more dense thermoplastic will sink, while the lighter one floats. If the fluid does not physically or chemically alter the thermoplastics, the separation should be extremely efficient. This process will be most efficient for mixtures of clean, dry pieces of thermoplastics.

Properties of Supercritical Fluids Relevant to the Separation of Recyclable Thermoplastics

The physical properties of compounds are strong functions of the system pressure {6,7} in the supercritical region, making them easily adjustable. For example, the density of a supercritical fluid can be isothermally adjusted from gas-like to liquid-like values via relatively small pressure changes without the occurrence of two-phase, gas-liquid equilibria. The density of a near-critical fluid (a liquid at a temperature just below the critical temperature and at a pressure at or above the vapor pressure) can be adjusted over a wide range of liquid-like values via relatively small pressure changes. The variable density of supercritical and near-critical fluids is the foremost property which makes them suitable for the separation of materials with exhibit very small density differences. Simply by varying the density of the fluid to a value which is intermediate to that of two thermoplastics, one will float and the other will sink.

The critical temperatures of commonly used fluids such as CO_2 are relatively mild, therefore the supercritical region is readily accessible and thermal degradation of the thermoplastics will not occur. The fluids chosen for use in this supercritical fluid process were carbon dioxide and sulfur hexafluoride. Both are non-toxic, non-flammable, commercially available on a large scale, and environmentally safe. Both have mild critical properties, listed in Table I. Carbon dioxide is the most commonly used supercritical fluid and can be compressed to densities in the range of 1000 kg/m^3. Since the separation of non-olefin thermoplastics will require fluid densities up to approximately 1400 kg/m^3, mixtures of carbon dioxide and sulfur hexafluoride, a very dense supercritical fluid, will be required.

The use of near-critical and supercritical fluids provides not only an **adjustable fluid density** at **mild temperatures,** but also these other advantages:

Low Viscosity - The settling rate of the separated material will be far greater in a supercritical fluid than in a conventional liquid due to the low viscosity of supercritical fluids.

Poor Solvency - Non-polar supercritical fluids such as CO_2 and SF_6 exhibit poor solvency with respect to most high molecular weight materials. Since separations are to be achieved by a sink-float method (not by dissolution or extraction), it is beneficial that little, if any, dissolution, swelling, or deformation of the waste thermoplastics occurs upon brief exposure to these fluids. This also implies, however, that the proposed process will be inefficient in the removal of additives or contaminants within the pieces of plastic, unless they are highly soluble in the fluid.

High Vapor Pressure Under Ambient Conditions - This property of the fluids insures that the thermoplastics separated in the proposed flotation process will dry instantaneously upon their removal from the process.

Minimal Environmental Impact - Use of a non-toxic, environmentally safe (no chlorinated solvents), non-flammable supercritical fluids in the recycling process eliminates one environmental problem without creating another. In fact, the fluid will be recycled through the process indefinitely, minimizing raw material costs.

Equation of State Model of Supercritical Fluid Density

An accurate and precise method of correlating and predicting CO_2/SF_6 mixture density data was developed. This was to provide a method of calculating the temperature, pressure and fluid composition needed to achieve a desired density. Equations of state are commonly used to relate pressure, temperature, density and fluid composition. Several types of equations were therefore evaluated, but the best results were obtained with a cubic equation of state which was designed specifically to provide accurate near-critical and supercritical fluid densities. The Lawal-Lake-Silberberg equation of state {8} has the following pressure-explicit form:

$$P = RT/(v-b) - a/(v^2 + \alpha\, bv - \beta b^2) \tag{1}$$

where

$$a = \Omega_a\, R^2 T_c^2 \gamma/P_c \tag{2}$$

$$b = \Omega_b\, RT_c/P_c \tag{3}$$

$$\Omega_a = [1 + (\Omega - 1)Z_c]^3 \tag{4}$$

$$\Omega_b = \Omega Z_c \tag{5}$$

$$\Omega = b/v_c \tag{6}$$

$$\alpha = (1 + (\Omega-3)\, Z_c)/(\Omega Z_c) \tag{7}$$

$$\beta = (Z_c^2(\Omega-1)^3 + 2\,\Omega^2\, Z_c + \Omega(1-3Z_c))/(\Omega^2 Z_c) \tag{8}$$

$$\gamma = [1 + \kappa\, (1 - T_r^{0.5})]^2 \tag{9}$$

$$\kappa = 0.14443 + 1.06624\omega + 0.02756\omega^2 - 0.18074\omega^3 \tag{10}$$

For mixtures, the following mixing rules should be used:

$$a = \Sigma\, \Sigma\, x_i\, x_j\, (1 - \delta_{1ij})(a_i a_j)^{0.5} \tag{11}$$

$$b = (\Sigma\, x_i\, b_i^{1/3})^3 \tag{12}$$

$$\alpha = \Sigma\, \Sigma\, x_i\, x_j\, (1 - \delta_{2ij})(\alpha_i \alpha_j)^{0.5} \tag{13}$$

$$\beta = \Sigma\, \Sigma\, x_i\, x_j\, (1 - \delta_{3ij})(\beta_i \beta_j)^{0.5} \tag{14}$$

The input parameters are:

T_{ci}, P_{ci}, Z_{ci}, ω_i, Ω_i, δ_{1ij}, δ_{2ij}, δ_{3ij} .

This particular equation of state provided excellent near-critical liquid density and supercritical fluid density correlations because of the following constraints applied to its use:

1. The critical temperature, pressure, and volume of each component were matched exactly.

2. The van der Waals covolume term, b, for each component was determined by optimizing the equation of state's correlation of pure component density data.

The input parameters required for the use of this equation of state for CO_2, SF_6, and mixtures thereof are provided in Table I.

TABLE I. CRITICAL PROPERTIES AND EQUATION OF STATE PARAMETERS FOR CARBON DIOXIDE AND SULFUR HEXAFLUORIDE

	T_c K	P_c MPa	Z_c	ω	Ω
CO_2	304.2	7.382	0.275	0.225	0.360
SF_6	318.7	3.759	0.281	0.286	0.359

$\delta_{1ij} = 0.31 \qquad \delta_{2ij} = 0.42 \qquad \delta_{3ij} = 0.18$

Experimental Method for the Determination of Fluid Density and Observation of Thermoplastic Separation

High pressure experiments were conducted using a JEFRI high pressure view cell (D.B. Robinson and Assoc.), which is shown in Figure 1. The core of this equipment is a heavy-walled quartz tube (interior volume of approximately 130 cm³) containing a short, moveable piston. The piston separates the supercritical fluid phase from the pressure-generating medium, a clear silicone oil. An O-ring held in place by Teflon spacers in the piston prevents the oil from contacting the fluid while permitting free movement of the piston. The tube and the silicone oil are encased in a windowed, 316 stainless steel vessel rated to 70 MPa. Pressure is generated within the quartz tube via injection of the silicone oil from a Ruska mechanized syringe pump, which moves the piston upwards, decreasing the volume occupied by the supercritical fluid.

The density of CO_2/SF_6 mixtures was measured using a number of calibrated ($\pm$ 0.0001) density floats from Techne, Inc., which were added to the upper chamber of the Robinson cell for each experiment. Fluid mixtures were prepared by weight in 300 cm³ lecture bottles equipped with Teflon stir bars and stirred for 15 minutes. The entire mixture (over 99%) was then charged to the Robinson cell using the Haskell gas booster. After allowing the temperature to equilibrate, the volume of the fluid mixture was slowly reduced, until each of the density-beads floated in sequence. Using this method, the pressure and temperature at which the fluid density was equivalent to the bead density could be determined to within 0.015 MPa and 0.1K, respectively.

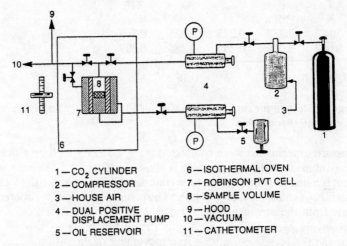

1 — CO$_2$ CYLINDER	6 — ISOTHERMAL OVEN
2 — COMPRESSOR	7 — ROBINSON PVT CELL
3 — HOUSE AIR	8 — SAMPLE VOLUME
4 — DUAL POSITIVE DISPLACEMENT PUMP	9 — HOOD
	10 — VACUUM
5 — OIL RESERVOIR	11 — CATHETOMETER

Figure 1. High pressure, variable volume view cell.

The Robinson cell was also used to determine the density, and therefore the potential for separation by density, of a number of commercially-available thermoplastics. In a typical experiment, thermoplastic mixtures (in the form of 0.3 cm x 0.3 cm squares) were charged to the quartz tube (above the piston) which was then sealed and filled with the fluid of choice. Following temperature equilibration, the pressure was gradually increased, and the various constituents of the thermoplastic mixtures were sequentially floated according to their densities. Following flotation of the entire mixture, the pressure was then slowly reduced to check the reversibility of the process and precision of the separation pressure measurements. The windows in the outer steel shell permitted visual observation of the entire process.

Research grade CO_2 and SF_6 were obtained as pressurized liquids from the Linde Division of Union Carbide Corporation. Post-consumer thermoplastic granule mixtures were received from Dow Chemical Western (Pittsburg, CA) and Econcorp (Pittsburgh, PA), and were used as received. The material from Dow had been previously separated by density using water, resulting in a "heavy" (PVC, PS, PET) and "light" (polyolefins) fraction. Material was also received from wire & cable and sources. These samples had previously been shredded (average size was approximately 1/8 - 1/4 inch in diameter) and had been passed through a water wash.

Results and Discussion

CO_2/SF_6 mixture density data measured using the calibrated density floats were fit to the Lawal-Lake-Silberberg equation of state. The best fit of mixture data was obtained using the interaction parameters noted in Table I. Some of these results are illustrated for near-critical liquid mixtures in Figure 2 and for supercritical fluid mixtures in Figure 3. Literature data {9-15} were used for the pure component data. Compositions are in weight fractions.

These figures also indicate the approximate densities of the six major recyclable thermoplastics, and the appropriate mixture required to achieve their separation. At supercritical conditions (about 333 K), for example, a fluid mixture containing about 85% SF_6 could be used to separate a mixture containing all of the thermoplastics since its density can be varied from 0 -1400 kg/m₃ as pressure is increased from 0 - 40 MPa. Pure carbon dioxide or CO_2/SF_6 mixtures containing less than 85% SF_6 could be used to separate the polyolefins (densities less than 1000 kg/m₃), while pure SF_6 or CO_2/SF_6 mixtures containing up to 15% CO_2 could be used to separate the non-olefin thermoplastics. Similarly, at ambient temperatures (about 295 K, Figure 2), pure CO_2 or CO_2/SF_6 mixtures containing up to 70% SF_6 could be used for the separation of polyolefins, while mixtures richer in SF_6 or pure SF_6 could be used to separate the more dense thermoplastics. Therefore, if the thermoplastic mixture was first separated into two mixtures using water as the flotation medium, the light cut (composed of thermoplastics with a density less than 1000 kg/m³) could be separated with pure CO_2 and the heavy cut (consisting of thermoplastics with a density greater than 1000 kg/m³) could be separated with pure SF_6.

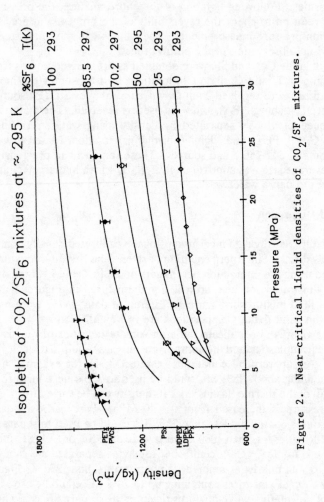

Figure 2. Near-critical liquid densities of CO_2/SF_6 mixtures.

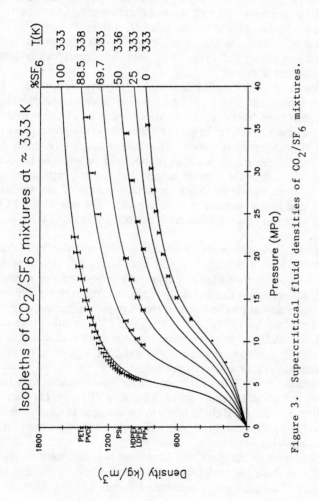

Figure 3. Supercritical fluid densities of CO_2/SF_6 mixtures.

A number of separations of thermoplastic mixtures were made using CO_2/SF_6 blends. The separation experiments indicated that any thermoplastics with different densities (even as small as several thousandths of a g/cm^3) could be separated. Their brief exposure to the high pressure fluid caused no discernable change in their appearance or properties. Particle densities were equivalent during the fluid compression and expansion cycles, indicating that no significant swelling occurred during this brief exposure (less than an hour) to the fluid environment. The equation of state was used to determine the density of each thermoplastic in the mixture, Tables II and III. This was accomplished by calculating the fluid density at the pressure at which the thermoplastic was 'suspended' in the fluid. Some important findings from these separations are noted below:

Given the density ranges exhibited by the polyolefins, pure CO_2, either as a liquid or supercritical fluid, can be used to separate polypropylene, LDPE, and HDPE into pure fractions. In addition, a degree of color separation is possible in the case of HDPE. In general, colorless (milk bottle grade) HDPE exhibits the lowest density, followed by the darker colors (green, black, blue), and finally the lighter colors, such as orange and yellow. Some overlap in density occurs amongst the various colors, and further experiments will be performed to more clearly establish the density ranges of the various colors commonly found in post-consumer HDPE waste. The size of the LDPE and PP fractions were too small to adequately address the issue of color separation for these materials.

Expanded polystyrene, while initially present with the "light" fraction of post-consumer waste, is densified rapidly and irreversibly in the presence of CO_2 from a bulk density less than 200 kg/m_3 to a bulk density of approximately 1050 kg/m^3, whereupon it can be easily separated from the polyolefins.

The transparent PVC commonly found in post-consumer waste exhibits a density lower than that of PET, while the filled PVC found in wire and cable scrap exhibits a wide range of densities, approx 1370 - 1580 kg/m^3. Thus significant density overlap between PVC and PET will only occur if post-consumer and wire/cable materials are mixed.

Separation of PET from post-consumer waste reveals four individual fractions, two transparent and two green. This occurs because (a) the green dye transesterified into the PET chain leads to an increase in density and (b) the neck and body of a PET bottle exhibit different densities, owing to different deformation histories.

The densities of the glass and aluminum particles were so high, 2400-2800 kg/m^3, that these impurities were separated simply by floating all of the thermoplastics. The glass and aluminum remained at the bottom of the cell under these conditions

Wood and paper impurities were also readily separated from the thermoplastics. Wood chips were the first particles to float as pressure was increased due to their low density of 400 - 800 kg/m^3. Paper products, such as writing paper, newspaper and glossy paper had densities of 1470, 1570 and 1650 kg/m^3, respectively, and therefore were separated from the glass and aluminum particles after the thermoplastics had been floated.

Future Work

The high-pressure apparatus illustrated in Figure 1 was used to demonstrate the feasibility of near-critical or supercritical fluids to separate thermoplastics by density. Although it was possible to observe the plastics separate in the high-pressure environment, it was not possible to extract the separated plastics from the apparatus to ambient pressure without the plastics mixing as they all sank to the bottom of the cell. In order to evaluate the purity of fractions removed from a near-critical or supercritical fluid environment, the unit in Figure 4 was constructed. This apparatus not only permits the recovery of thermoplastics from the high-pressure environment of the stirred autoclave via the pressure trap below the autoclave, but also allows for the recycle of the supercritical fluid. Only a small amount of low-pressure gas will be lost when the plastics are discharged from the pressure trap.

The operation of this apparatus will result in longer residence times in the supercritical fluid and higher ratios of fluid mass/particle mass, increasing the possibility of swelling and extraction of CO_2-soluble contaminants. Therefore, the conditions at which separations should occur will not only be predicted using the equation of state (and assuming no swelling), but also monitored visually in a visual cell in case a slight amount of swelling does occur.

Table II
Density Results for Various Thermoplastic Materials
Polyolefins

	kg/m³
HDPE - Post-consumer waste	
- Milk Bottle grade (transparent)	950 - 955
- Dark colors (black, green, blue)	955 - 965
- Light Colors (yellow, orange)	970 or more
HDPE - Wire & Cable	
- Various colors	950 - 960
LDPE - Post-consumer waste	
- Dark colors	930 - 935
LDPE - Wire & Cable	
- Black	940 - 950
- Other colors	930 - 940
PP - Post-consumer waste	
- White	920 and lower
Ethylene/Hexene Copolymer - mixed with HDPE	
- Blue	950

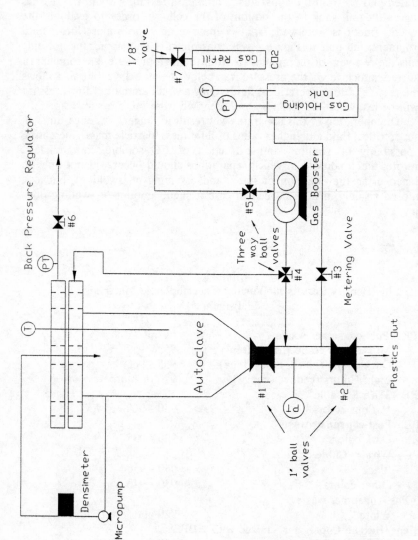

Figure 4. High pressure apparatus which permits removal of separated plastics from high pressure environment and recycle of the fluid.

Table III
Density Results for Various Thermoplastic Materials
Non-Olefins

	kg/m³
Polystyrene - Post-consumer	
- Densified styrofoam	1050
- Clear	1100
- Dark colors (blue, black)	1160 - 1220
- light colors	1220 and higher
PVC - Post-consumer	
- Clear, bottle grade	1320 - 1330
- Filled (from pipe)	1500 - 1550
PVC - Wire & Cable	
- Colorless	1370 - 1380
- Light colors (beige)	1370 - 1480
- Medium Colors (red)	1480 - 1510
- dark colors	1510 - 1580
PET - Post-consumer	
- Clear - Neck	1330 - 1335
Body	1350 - 1360
- Green - Neck	1340 - 1350
Body	1360 - 1365
Fluoropolymers - Wire & Cable	1550 and higher

This instrument will also be used to evaluate the effect of loading (% solids) on overall separation efficiency (ie., the importance of hindered settling). From preliminary results, it appears that hindered settling does not become problematic until the solids loading approaches and exceeds 20%.

Conclusions

Pure carbon dioxide was used to separate polyolefin thermoplastics and pure sulfur hexafluoride was to separate the non-olefin thermoplastics. Carbon dioxide/sulfur hexafluoride blends which were capable of separating all of the thermoplastics, such as a mixture of 85% SF_6 and 15% CO_2 at 333K, were also prepared.

Wood, paper, glass and aluminum impurities were easily separated from the thermoplastics.

Separations of thermoplastics with extremely similar densities, such as PVC from PET, an ethylene-hexene copolymer from polyethylene, and blue polystyrene from untinted were achieved.

The separations were attained at mild temperatures (293K - 333K) and moderately high pressures of 2 - 40 MPa. Exposure to the fluids did not appear to cause any physical or chemical change to the thermoplastic, and the samples 'dried' instantly upon depressurization. It is unlikely, however, that this process will effectively remove most additives or contaminants found within the pieces of plastic. The separations will also be most efficient with washed mixtures, since large amounts of contaminants on the particle surfaces will alter the particle density.

The Lawal-Lake-Silberberg equation of state was able to provide an excellent correlation of near-critical and supercritical fluid density.

An automated batch separation unit capable of discharging the separated samples from the high-pressure environment has been built. This apparatus will provide a means of evaluating the separation efficiency of the process.

Acknowledgments

We would like to thank the Plastics Recycling Foundation and the University of Pittsburgh Central Research and Development Fund and the Texaco Foundation for their generous support of this project.

Literature Cited

1. Thayer, A.M. *Chem. Eng. News* **Jan. 30, 1989**, p. 7
2. Voss, D. *Chem. Eng. Prog.* **Oct. 1989**, p. 67
3. Mackzo, J. *Plast. Eng.* **April 1990**, p. 51
4. Stolzenberg, A. *Proceedings of Recyclingplas IV*, **1989**, p. 129
5. Martin, A.E. *Small-Scale Resource Recovery Systems*, Noyes Data Corp., Park Ridge, NJ, **1982**
6. McHugh, M.A.; Krukonis, V.J. *Supercritical Fluid Extraction*, Butterworths, Stoneham, MA, **1986**
7. Paulaitis, M.E.; Krukonis, V.J.; Kurnik, R.; Reid, R. *Chem. Eng. Rev.* **1982**, 1, p. 174
8. Lawal, A.S.L. *Improved Fluid Property Predictors for Reservoir Compositional Simulation*, Ph.D., Dissertation, The University of Texas at Austin, **1984**
9. Biswas, S.N.; ten Seldam, C.A. *Fluid. Ph. Equil.* **1989**, *47,* p. 67
10. Blanke, W.; Haeusler, H.; Weiss, R. *Int. J. Thermophys.* **1988**, *9*, p. 791
11. Prisyazhnyi, A.P.; Totskii, E.E. *Teplofiz. Vys. Temp.* **1987**, *25*, p. 887
12. Watanabe, K.; Watanabe, H.; Oguchi, K. *Proc. Symp. Thermophys. Prop.* **1977**, *7*, p. 489
13. Ulybin, S.A.; Zherdov, E.P. *Dokl. Akad. Nauk. USSR* **1970**, *191*, p. 572
14. Biswas, S.N.; Trappeniers, N.J.; Hoogland, J.H.B. *Physica A* (Amsterdam) **1984**, *126A*, p. 384
15. Angus, S.; Armstrong, B.; de Reuck, K.M. *Carbon Dioxide*, Pergamon Press, NY, **1976**

RECEIVED April 28, 1992

Chapter 14

Recoverable Polyimide Copolymers

Yulong Wu, Giuliana Tesoro, and Israel Engelberg[1]

Department of Chemistry, Polytechnic University, Brooklyn, NY 11201

Experimental copolymers of Dithiodianiline bismaleimide with aromatic diamines and with o,o'-Diallylbisphenol A have been prepared. Properties comparable to those of a commercial bismaleimide resin (Matrimid 5292) have been obtained for some copolymers. Solubilization of cured copolymers by reduction of disulfide bonds is feasible, and re-curing (resetting) can be carried out by reaction of thiol groups formed by reduction with polyfunctional reagents to yield modified resins of interesting properties.

In our continuing research on thermosets designed for post-setting recovery of value (tertiary recycling), we have shown that epoxy resins cured with disulfide-containing crosslinking agents can be reduced to the point of complete solubilization, and reset by oxidation or by chemical modification to provide thermosets of satisfactory properties (1-4). Subsequent work has shown that the concepts can, in principle, be extended to recoverable polyimides containing disulfide bonds (5,6). For example, homopolymers of 4,4'-Dithiodianiline bismaleimide (DTDA-BM) could be reduced to the point of complete solubilization after curing (5). However, it was postulated that resin properties, solubilization by reduction, and resetting of reduced products would be more easily controlled in the case of copolymers of disulfide-containing monomers such as DTDA-BM, where the structure of the cured resin, the disulfide content and the crosslink density could be varied.

The objective of the present study was to prepare and evaluate selected copolymers of DTDA-BM with aromatic diamines, and with o,o'-Diallylbisphenol A (DABPA) and to determine whether thermal and mechanical properties

---

[1]Current address: Rafael, P.O. Box 2250, Haifa 31021, Israel

NOTE: This chapter is part III in a series.

comparable to those of a commercial bismaleimide resin could be attained, while maintaining the feasibility of recovering cured polymers by reduction. In the work reported here, properties of experimental copolymers of DTDA-BM have been compared with those of a commercial two-component system containing 4,4'-Methylenedianiline bismaleimide (MDA-BM) and DABPA (Ciba Geigy, Matrimid 5292-components A and B), for which thermal and dynamic mechanical properties have been reported (7,8). After reduction and solubilization of cured copolymers, approaches to resetting were explored for selected systems.

Experimental

Materials and Methods. The monomers used are shown in Figure 1. DTDA-BM was prepared as reported previously (5). Matrimid 5292 (components A and B) was obtained from Ciba Geigy and used as received. All other materials (from Aldrich) were used as received.

Thermal properties were determined by DSC (DuPont 910) and TGA (DuPont 951) under nitrogen at heating rates of 10°C/min and 20°C/min respectively. Dynamic mechanical analysis was carried out in air (DuPont 982) at a heating rate of 5°C/min.

Preparation of Copolymers and Curing. Copolymers of DTDA-BM with aromatic diamines have been previously investigated by DSC (6) with the objective of identifying preferred diamine structures and monomer ratios for copolymer synthesis.

For the present study, copolymers of DTDA-BM with 4,4'-Diaminodiphenyl sulphone (DDS) were prepared by a modification of the procedure of Crivello (9) at DDS/DTDA-BM mol ratios varying from 0.3 to 0.5. Monomers were mixed at 190°C, stirred until clear, cast and cured in the mold as previously reported (6).

Copolymers of DTDA-BM and of MDA-BM with DABPA were prepared at mol ratios of 1/1. The monomers were mixed, heated to 160-180°C with stirring until homogeneous, and degassed in a vacuum oven (20 minutes at 150°C) prior to curing. For all copolymers studied, curing conditions were determined by DSC and are as indicated in the tables.

Results and Discussion

Polymer Properties. Salient thermal and mechanical properties of homopolymer from DTDA-BM are compared with those of MDA-BM in Table I and Figure 2. The lower thermal stability of the DTDA-BM homopolymer, attributed to the presence of the disulfide bonds, is evident from the decomposition temperature (TGA). A more precise evaluation of thermal response requires consideration of the effect of curing conditions on the properties of the cured polymers.

The results of the DSC study reported previously (6) for copolymers of DTDA-BM with the aromatic diamines shown in Figure 3 formed the basis for the selection of DDS as the preferred diamine, and for mol ratios of DDS to

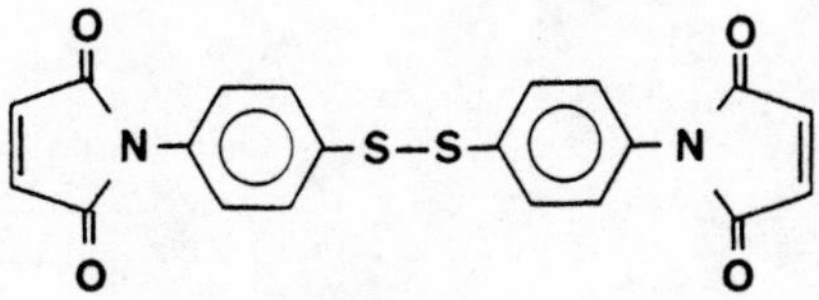

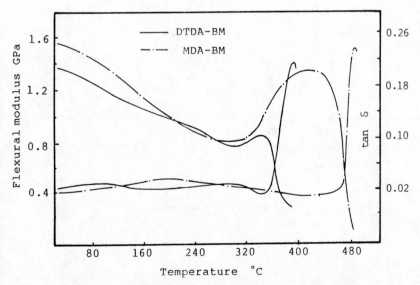

4,4'-Dithiodianiline bismaleimide
(DTDA-BM)

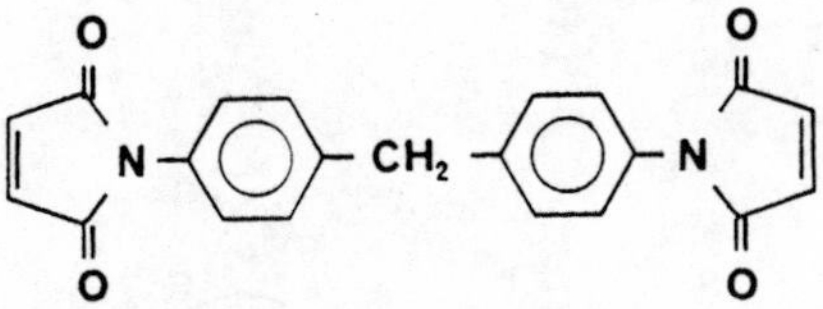

4,4'-Methylenedianiline bismaleimide
(MDA-BM)

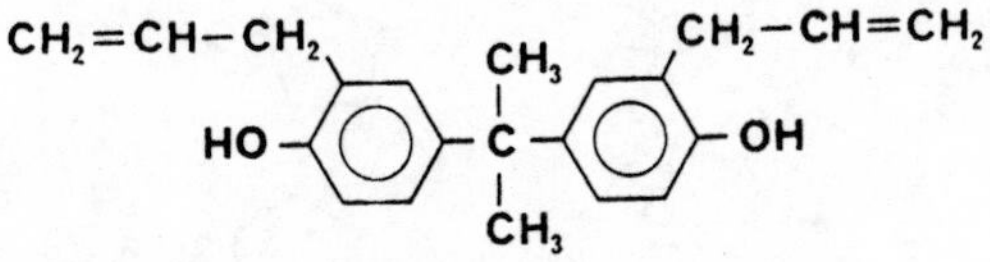

o,o'-Diallylbisphenol A
(DABPA)

Figure 1.   Monomers

Figure 2   DMA of neat DTDA-BM and MDA-BM

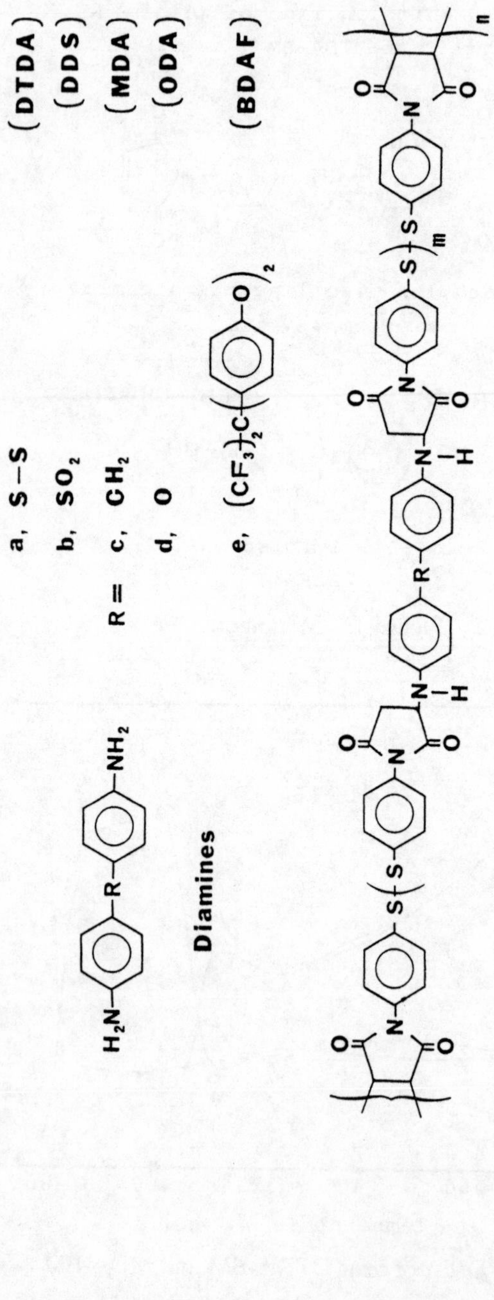

Figure 3. Copolymers of DTDA-BM with aromatic diamines

Table I. T_g, T_d and E_f of DTDA-BM and MDA-BM Homopolymers

Bismaleimide	T_g °C (DSC)	T_d °C (TGA)	E_f^a GPa (DMA)
DTDA-BM[b]	334	353	1.30
MDA-BM[c]	335	519	1.50

[a] E_f at 50°C

[b] Samples cured at 190°C/4 hr + 220°C/4 hr + 250°C/4 hr

[c] Samples cured at 170°C/4 hr + 200°C/4 hr + 250°C/4 hr

Table II. Properties of Cured Copolymers[a]
of DDS/DTDA-BM at Varying Mol Ratios
A. Thermal Properties

Diamine	Mol ratio	T_g °C (DSC)	T_g °C (DMA)[b]	(TGA) T_d °C	(TGA) R^c %
	0.3	284	280	360	62.7
DDS	0.4	266	278	354	64.5
	0.5	260	275	347	60.3
DTDA-BM (neat)		334	358	353	68.1

[a] All samples are cured at 190°C/4 hr + 220°C/4 hr + 250°C/4 hr.

[b] The initial transition temperature of tan δ peak is considered as T_g.

[c] Residue at 550°C

B. Mechanical Properties

DDS/DTDA-BM (by mol)	E_f GPa (DMA) °C 50	150	200	250	300	350
0.3	2.15	1.82	1.65	1.40	0.89	-
0.4	1.69	1.40	1.26	1.08	0.55	-
0.5	1.34	1.12	1.02	0.875	0.31	-
DTDA-BM (neat)	1.30	1.06	0.961	0.863	0.75	0.827

DTDA-BM ranging from 0.3-0.5. DSC traces for these copolymers are shown in Figure 4. The higher polymerization temperature of the copolymers, and the corresponding increase in the processing window as compared to the DTDA-BM homopolymer are clearly shown. Thermal and mechanical properties of the copolymers are summarized in Tables IIA and IIB. In spite of the lower concentration of disulfide bonds, the thermal stability of the copolymer is not improved as compared to that of the DTDA-BM homopolymer (Table IIA). The improvement resulting from lower disulfide content may be offset by a decrease in crosslink density in the case of the copolymers. It has been shown that properties of cured copolymers of DTDA-BM with aromatic diamines depend in part on diamine structure and mole ratio (6). Further work is needed to identify optimum compositions and curing conditions for copolymers that may be of interest for specific applications.

For cured copolymers of DTDA-BM and of MDA-BM with DABPA at 1/1 mol ratio, thermal properties are summarrized in Table III. Dynamic

Table III. Thermal Properties of Cured Bismaleimides/DABPA Copolymers

Bismaleimde/DABPA =1/1 (by mol)	Curing condition °C/hrs	T_g °C		T_d °C
		DSC	DMA	TGA
DTDA-BM/DABPA	200/12	254	240	402
MDA-BM/DABPA[*]		226	244	433
DTDA-BM/DABPA	200/6 +250/6	276	291	412
MDA-BM/DABPA[*]		319	319	441

[*] MDA-BM/DABPA--Matrimid 5292 (Ciba-Geigy)

mechanical responses (flexural modulus E_f and tan δ) for cured copolymers of DTDA-BM and of MDA-BM with DABPA are compared in Table IV and Figure 5.

Predictably, the onset of decomposition temperature for the DTDA-BM copolymers is lower than for the disulfide-free MDA-BM copolymers (Matrimid 5292). A comparison of the thermal response of the DTDA-BM/DABPA copolymers (Table III) with that of the DTDA-BM homopolymer (Table I) is of interest. The onset of decomposition temperature (T_d) is higher in the copolymer, where the concentration of disulfide is lower and crosslink density is presumably comparable. The lower T_g of the copolymer (276°C as compared to

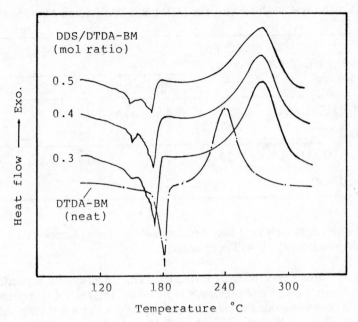

Figure 4.   DSC of curing for DTDA-BM and for DTDA-BM/DDS
           copolymers

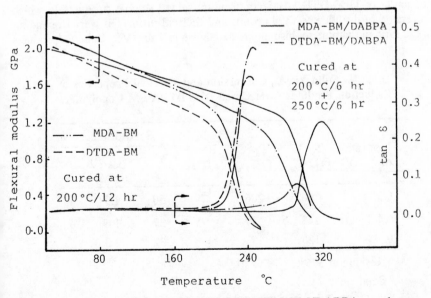

Figure 5.   DMA of MDA-BM/DABPA and DTDA-BM/DABPA copolymers

Table IV. Flexural Modulus (E_f) For DTDA-BM/DABPA and Matrimid 5292

Bismaleimide/DABPA = 1/1 (by mol)	Curing condition °C/hrs	E_f GPa (DMA) °C				
		50	100	200	275	325
DTDA-BM/DABPA		1.93	1.66	1.14	-	-
Matrimid 5292	200/12	1.95	1.78	1.39	-	-
DTDA-BM/DABPA	200/6 +	2.06	1.83	1.43	0.773	-
Matrimid 5292	250/6	2.07	1.86	1.53	1.25	0.123

334°C for the homopolymer) may be attributed to the presence of aliphatic groups derived from the DABPA comonomer.

Recovery of Cured Copolymers by Reduction. The conditions for reduction of cured copolymers were those previously reported for DTDA-BM homopolymer (5). Cured resin was ground to 600 micron size particles. To a one gram sample was added 50 ml of diglyme, 17 drops of dilute HCl (from 15 ml of water containing 2 drops of concentrated HCl) and 7 ml of tributyl phosphine. The mixture was refluxed under nitrogen until complete solubilization was attained. The reduced product (precipitated in methanol) was filtered and dried. Reduction of DDS/DTDA-BM copolymers yielded soluble products in which conversion of disulfide to thiol groups and thermal properties were dependent on the ratio of diamine to bismaleimide (shown in Table V). Soluble products

Table V. Solid Yields, Conversion and Thermal Properties of Reduction Products from Cured DDS/DTDA-BM Copolymers

Mol ratio of DDS	T_g °C (DSC)	T_d °C (TGA)	R^a % (TGA)	Solid products		
				Y^b %	-SH(10) meq/g	X %
0.3	166	206	79	50	1.9	46
0.4	142	179	75	37	1.6	40
0.5	126	153	68	32	2.0	53

[a] Residue at 375°C

[b] Solid yield

[c] % Conversion of -S-S- to -SH

obtained by reduction of DTDA-BM/DABPA copolymers have not been explored in depth.

The products obtained by reduction of DTDA-BM copolymers can be reset by curing in the presence of polyfunctional reagents that can react with thiol groups. In one approach, the reduction product from the cured copolymer of DTDA-BM/DDS (1/0.5 mol ratio) was reset by curing with epoxy resin (Epon 828) yielding thermoset interpenetrating networks in which curing behavior and resin properties were dependent on the ratio of thiol to epoxy functional groups (6). It was shown that thiol groups in the reduced products react readily with added epoxide: With Epon 828 as the epoxy resin, reset was initiated at 80-100°C and completed at 160-200°C, reflecting a good curing profile. The T_g of the reset modified polymer was comparable to that of epoxy resin cured with an aromatic diamine. These results provide an example of an approach to the utilization of the recovered (reduced)copolymers for the design of modified resins. Further study of the chemical modification of reduced polyimide copolymers for improved properties merits consideration.

Conclusions

The disulfide-containing monomer 4,4'-Dithiodianiline bismaleimide (DTDA-BM) is a starting material for the synthesis of recoverable polyimide copolymer resins that exhibit thermal and mechanical properties of interest for many applications.

Copolymers with aromatic diamines afford a broad range of products where processing and performance properties can be varied depending on the the diamine structure and mole ratio used.

Properties of copolymers of DTDA-BM with o,o'-Diallylbisphenol A (DABPA) compare favorably with those of a commercially available MDA-BM/DABPA copolymer system (Matrimid 5292, Ciba Geigy).

Disulfide bonds in the cured crosslinked copolymers can be cleaved by reduction, yielding soluble products containing thiol groups. The reduced products can be recovered for re-processing, reset, or modified by curing in the presence of thiol-reactive polyfunctional reagents to yield thermosets of improved properties.

Acknowledgments. The authors gratefully acknowledge financial support for this work by the U.S. Department of Energy, Advanced Industrial Concepts Division, Materials Program.

Literature Cited.

1. Tesoro, G.; Sastri, V. *Makromol. Chem., Macromol. Symp.* **1989,** *25,* 75-84.
2. Tesoro, G.; Sastri, V. *J. Appl. Polym. Sci.* **1990,** *39,* 1425-1438; 1439-1458.
3. Engelberg, P. I.; Tesoro, G. *Polym. Eng. Sci.* **1990,** *30,* 303-307.
4. Tesoro, G.; Sastri, V. *USP* 4,882,399 **1989,**
5. Sastri, V.; Tesoro, G. *J. Polym. Sci., Part C,* **1989,** 153-159.

6. Wu, Yulong; Tesoro, G. *Polym. Adv. Technol.* **1990,** *1,* 253-261.
7. Zahir, S. A.; Renner, A. *USP* 4,100,140 **1978,**
8. Chattha, M. S.; Dickie, R. A. *J. Appl. Polym. Sci.* **1990,** *40,* 411-416
9. Crivello, J. V.; *J. Polym. Sci. (Chem.)* **1973,** 1185-1200.
10. Bald, E. *Talanta* **1980,** 281-282

RECEIVED March 24, 1992

Chapter 15

Incorporating Postconsumer Recycled Poly(ethylene terephthalate)

A New Polyester Resin

R. E. Richard, W. H. Boon, M. L. Martin-Shultz, and E. A. Sisson

The Goodyear Tire & Rubber Company, Akron, OH 44305

A new process is described which can be used to incorporate post-consumer PET (pcPET) into a bottle grade polymer. The process involves depolymerization of pcPET using ethylene glycol, purification of the resulting oligomeric products, and repolymerization in the presence of virgin raw materials. Using this process, PET polymers containing 10 and 20 weight % pcPET were prepared. The polymers had properties comparable to those of virgin bottle polymer. Using this technology, Goodyear has developed a new product, REPETE polyester resin, which is used to make bottles for both non-food and food-grade applications.

In 1989 approximately 700 million pounds of poly(ethylene terephthalate) (PET) were consumed to produce soft drink bottles (*1*). Presently about 28% of this material is recycled with the remainder being placed in landfills or incinerated. Due to problems with overloaded landfills and the negative environmental image plastics have received recently, legislation aimed at establishing bottle deposits and curbside recycling programs has increased tremendously. The collection of PET bottles through these programs has resulted in a source of post-consumer PET (pcPET) which has been used for a number of applications. Examples of such applications are polyols for unsaturated polyesters or polyurethanes, fiberfill, carpet, strapping and non-food grade bottles. Recycled PET is also blended with other materials such as poly(butylene terephthalate) (PBT), polycarbonate, or glass fibers, etc, for automotive as well as other engineering applications (*2*). Even before the commercialization of the PET soft drink container, Goodyear developed a process to recover PET from the diverse components of the bottle (*3,4*). This technology was used to construct a pilot plant demonstrating the recycling potential of PET soft drink bottles. The pilot plant is now located at the Rutgers University Plastics Recycling Center. Current commercial processes used to separate the components of PET bottles are based on the principles of this technology.

Goodyear recently developed a commercial product trademarked REPETE polyester resin, which incorporates recycled PET along with virgin starting material

0097–6156/92/0513–0196$06.00/0
© 1992 American Chemical Society

to make bottle grade polymer. REPETE polyester resin can be used for the injection blow molding of oriented containers for use in both non-food and food-grade packaging. This paper discusses the technical hurdles involved with incorporating pcPET into a commercial product for bottle applications.

Results and Discussion

The reuse of pcPET involves a number of important considerations including collection, cleaning, separation of non-PET materials, and reprocessing. In all phases of the recycling process a key issue is the purity and safety of the final product. Other issues that need to be addressed are removal of foreign materials, as well as the physical and chemical properties of the final product.

Cleaning, Collection, and Separation. The two commercially significant sources of pcPET are soda bottles collected through community curbside collection and bottle deposit return programs. In curbside collection programs, both PET bottles as well as other recyclable containers (aluminum and steel cans, glass, etc.) are collected and sorted by material type. The bottles are then either ground or baled and sent to a recycler. The pcPET obtained from bottle deposits is presorted at the point of collection and usually baled and sent directly to the recycler. The recycler is responsible for converting the bottles into clean PET flake that meets the specifications established by the enduser. The steps taken to ensure a pure product include a pre-cleaning and preparation, cleaning, and quality assessment.

The pre-cleaning procedure depends on whether the recycler receives the bottles in baled or ground form. In the case of baled bottles the non-PET bottles are removed, and the bottles may be separated by color depending on the enduse. The bottles are also inspected for contamination and then converted to flake. Ground bottles require more supplier education than baled bottles because once the bottles are converted to flake, sorting mistakes are hard to correct. The recycler receiving the ground bottles generally provides the collector with the grinder and establishes operating procedures that must be followed. The cleaning process consists of five steps which are:

> LABEL REMOVAL
> FLOAT SEPARATION
> HOT CAUSTIC WASH
> RINSE AND FLOAT SEPARATION
> ALUMINUM REMOVAL

The cleaning process starts when ground bottles are subjected to positive and negative air flow to remove paper and polypropylene film which are associated with the labels. The process also removes fines created during the grinding.

The ground flake is next suspended in a water float tank at ambient temperature to hydraulically separate other plastics from the PET using the density differences of the components. Using this technique, PET with a specific gravity of 1.30-1.36 g/cc will sink while the other polymeric components of the bottle such as high density polyethylene (HDPE), polypropylene (PP) or ethylene-vinylacetate (EVA) will float.

Subsequent to the plastics removal, the PET is suspended in a hot solution of caustic and surfactant to remove the ethylene-vinylalcohol (EVOH) and EVA based adhesives which are used to adhere both the label and the HDPE base cup to the bottle. The flake is then resuspended in water to further remove caustic and any remaining non-PET plastics . The last step in the cleaning process is the separation of aluminum from the PET. This is generally accomplished using an electrostatic separation. This technique involves subjecting the mixture to an electric field which remains on the

PET much longer than on the aluminum. As a consequence, the PET will adhere to a rotating wheel separating device, while the aluminum is removed by centrifugal force.

The resulting PET is then subjected to a quality assessment to certify compliance with the purchasing specification. The test measures the flake for aluminum, HDPE, PP, PVC, paper, wood, and green PET. The current flake specification in place to prepare REPETE polyester resin is shown in Table I

Table I. Goodyear Flake Specification for pcPET

Contaminant	Maximum [a]
Water	1 %
Green PET	1000 ppm
EVA	10 ppm
HDPE, PP	20 ppm
Aluminum	1 ppm
PVC	nil

[a]Concentrations on a weight basis

The flexibility of the specification is highly dependent on the ability of the reprocessing system to remove a given impurity. The specification for the flake used to prepare REPETE polyester resin is evolving as the effectiveness of Goodyear's reprocessing system is optimized.

One of the most challenging aspects associated with the incorporation of pcPET back into commercial bottle grade polymer stems from the chemically heterogeneous nature of the pcPET recycle stream. This heterogeneity results from differences in the technology used by various manufacturers of PET, as well as from different components which make up a PET bottle. The two commercially important routes to produce PET involve either the direct esterification of terephthalic acid (TA) with ethylene glycol (EG), or the transesterification of dimethylterephthalate (DMT) with EG, followed by polycondensation. While both processes produce the same basic polymer, variations in the amount and type of metal catalysts used in the processes gives rise to a mixture of catalytic materials in the pcPET recycle stream. Table II shows the various metals present and their probable function in a representative sample of pcPET.

Table II. Catalysts and Additives in pcPET Sample

Metal	Conc. (ppm)	Function
Sb	220-240	Polycondensation
Co	50-100	Polycondensation
P	6-110	Stabilizer
Mn	20-60	Transesterification
Ti	0-80	Polycondensation
Fe	0-6	Intro. during washing
Na, Mg, Si	Trace	Various

A second source of chemical heterogeneity comes from other monomers which are added to modify the properties of PET for specific end uses. These comonomers are used to control properties such as melting point, Tg, crystallinity, or rate of crystallization. In addition, chemical reaction during polymerization also results in the formation of diethylene glycol (DEG) through acid catalyzed etherification of EG under the conditions of polymerization. Table III shows examples of these comonomers and the amounts detected in a typical sample of pcPET.

Table III. Monomers Detected in pcPET (Excluding TA and EG)

Monomer	Weight%
DEG	1.8-2.2
Isophthalic acid	0-0.7
Cyclohexanedimethanol	0-0.3

A third source of heterogeneity in pcPET results from non-polyester materials which are typically found in the recycle stream. This issue is usually handled by the collection and recycling process as discussed earlier. In addition, the processing technology used to produce products from pcPET may also be designed to handle variations in the pcPET stream.

Another consideration which must be addressed is the initial properties of the pcPET as received from the recycler. Table IV compares some important properties of representative samples of pcPET with those of virgin PET bottle polymer. The differences in pcPET properties compared to unprocessed virgin PET are attributed primarily to the additional thermal history experienced by the pcPET. This results in decreased molecular weight (lower IV and increased COOH concentration) and an increase in both color and acetaldehyde level. The rigorous conditions of cleaning the pcPET may also contribute to the property differences.

Table IV. Properties of pcPET Compared with Virgin Bottle Polymer

Property	pcPET	Virgin PET
IV (dl/g)	0.46-0.76	0.72-0.84
Color b (yellowness)	0.4-4.0	0.0-1.5
Tm (°C)	247-253	244-254
COOH (eq/10^6 g)	22-32	11-26
Acetaldehyde (ppm)	9.9-10.7	1.2-5.5

Reprocessing Options for pcPET. A variety of technologies are available for the reprocessing of pcPET into marketable products. Some of the more commercially feasible options are:

pcPET "AS IS" (NO REPROCESSING)
EXTRUSION/REPELLETIZATION
GLYCOLYSIS/REPOLYMERIZATION
REVERSION TO RAW MATERIALS/REPOLYMERIZATION

The use of pcPET without additional reprocessing has been applied to the production of materials which do not demand the full complement of properties available from PET. This includes applications such as fiberfill and industrial strapping.

At the opposite end of the spectrum, pcPET may be reverted back to its raw materials under the appropriate conditions. This is possible due to the equilibrium nature of the reaction used to synthesize PET. The process used to regenerate the starting materials involves a solvolysis reaction, which determines the nature of the products obtained. The three most common solvolytic reactions used for this purpose are:

METHANOLYSIS
HYDROLYSIS
GLYCOLYSIS

The methanolysis process involves the complete depolymerization of the pcPET to yield DMT and EG (5). The DMT can then be subjected to distillation and recrystallization to produce polymer grade DMT. The EG is generally sold for other purposes such as antifreeze. Similarly, hydrolysis also can result in total depolymerization of the pcPET using water, often in conjunction with an acid or base catalyst, producing TA and EG (6). This process is less favorable due to the difficulties associated with the purification of the TA. The glycolysis process differs slightly from the previous two processes in that it does not lead to the formation of a discrete chemical species, but rather gives bis(2-hydroxyethyl) terephthalate (BISHET) along with higher oligomers (7,8). The relative amounts of BISHET and oligomers produced depends on the amount of EG added and the conditions of the reaction.

In between these extremes there exists a number of other options available for the reprocessing of PET. One reprocessing technology used commercially involves the melt extrusion/filtration of pcPET followed by repelletization. This process provides a degree of purification by removing unmeltable impurities by filtration and can be used to prepare polymer which can be converted into good quality bottles by injection-blow molding, for non-food applications.

The glycolysis/repolymerization process is closely related to the process involving reversion to raw material described earlier, in that it involves solvolysis with EG. This process has been integrated by Goodyear into a proprietary technology which can depolymerize pcPET, purify it using state of the art processing techniques, and repolymerize it in the presence of virgin TA and EG. Using this technology, Goodyear has developed REPETE polyester resin which can incorporate various levels of pcPET along with the virgin starting materials.

The merits and shortcomings of the above processes need to be examined as to their economic feasibility. This evaluation is outlined in Figure 1. Where no reprocessing is used, one would expect relatively low purity and uniformity, with no additional reprocessing cost. The potential value of the pcPET would be that of finished resin, but at lower cost. At the other end of the spectrum, reversion of pcPET to raw materials results in high purity and uniformity, however the associated reprocessing costs are reported to be high (9). This approach utilizes the raw materials value of the pcPET. In the case of extrusion/repelletization, one can improve the purity of the pcPET to some extent, although additional reprocessing costs are incurred. As in the first case, the potential value of the pcPET approaches finished resin value. Finally, considering the glycolysis/repolymerization process used in the manufacture of REPETE polyester resin, Goodyear is able to maximize the purity and uniformity of the polymer, as well as the potential value of the pcPET. This process also minimizes the associated reprocessing costs to obtain the final product.

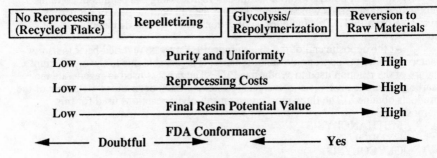

Figure 1. Reprocessing Options For pcPET

One of the most challenging problems facing the producers of PET is the development of processes which are capable of incorporating pcPET into food grade PET. Figure 1 shows the evaluation of the probability of each of the previously discussed processes in achieving this goal. The probability of producing food grade PET with either no reprocessing, or with extrusion/filtratation is expected to be low. In the case of glycolysis, the FDA has recently given approval for this process. Similarly, the FDA has given approval to the process where methanolysis is used for the regeneration to raw materials (*10*). The glycolysis/repolymerization process has undergone extensive evaluation by Goodyear. This evaluation was designed to assure that PET containing pcPET was identical to virgin PET in purity, uniformity and processing characteristics. In addition, validation studies have been performed in which the pcPET was intentionally spiked with potential consumer misuse products in order to confirm their complete removal.

Contaminant Removal. From a health and safety perspective, any process which utilizes pcPET must be capable of removing potential contaminants that may be present in the recycle stream. The major contamination sources focused on by Goodyear in designing the REPETE polyester resin process are:

MICROBIOLOGICAL CONTAMINATION
FOREIGN MATERIALS
NON-PET PLASTICS
CONSUMER MISUSE CONTAMINANTS

The glycolysis/repolymerization process used by Goodyear in the manufacture of REPETE polyester resin, in combination with prudent separation practices by the recycler has led to an effective route for dealing with these issues. Microbiological contamination can result from incomplete washing of post-consumer bottles, leading to microbic growth. The caustic wash carried out by the recycler helps eliminate microbic contamination. This treatment is at higher temperatures and for longer time than that used in the washing process for returnable/refillable bottles. In addition, the high temperatures and pressure associated with the REPETE polyester resin process are more harsh than an autoclave and would be expected to eliminate any microbes that were present.

The types of foreign materials that may be found in pcPET include glass, wood, aluminum, and fibers (from paper, cotton, etc.). These materials are most effectively removed during the sorting and cleaning of the pcPET as discussed earlier.

Non-PET plastics are also handled primarily at the recycler level. This is carried out using density separation with water as the medium. The one plastic material that is not effectively removed by this method is PVC, due to its similar density to PET. Goodyear has recently addressed this problem by developing a unique technology to effectively remove PVC. This technology is based on selective surfactant froth flotation which is widely used to separate ores by the mining industry. The Mitsui Mining & Smelting Company had done much work in the 1970's applying this technology to the plastics industry. An example of their research has proven that PP can be separated from PE (11).

Consumer misuse contaminants would include materials such as gasoline, motor oil, petroleum products, and pesticides. These contaminants would be removed to a great extent during the recycler's flake washing process. Any contaminant that is absorbed into the bottle would be more difficult to remove by washing, and would be dealt with during the glycolysis/repolymerization process. Since this process involves the reduction of the high molecular weight PET to low molecular weight oligomers, it would unlock any absorbed materials from the plastic and facilitate its removal under the conditions of the process.

In order to validate the removal of consumer misuse contaminants, experiments were conducted which simulated the possible routes and degree of contamination of pcPET. The experiments consisted of intentional contamination of 2 liter PET containers using representative contaminants, including pesticides, herbicides, and petroleum based products which a consumer might store in PET food containers. The following criteria was used to select representative compounds for potential contaminants:

1. Banned products were eliminated since they are unlikely to enter the recycle stream
2. By-products of contaminant degradation in the environment were eliminated.
3. Products that are inaccessible to consumers, such as compounds used only by professionals, were eliminated.
4. Products applied as aerosol or dry powders were eliminated since those diluted prior to use have the highest probability of being present in pcPET.
5. The probability of developing analytical tests to detect the selected compounds at levels commensurate with their toxicity was necessary.
6. The compounds selected should have a high probability of being absorbed into PET.
7. A compound was selected from each of the four classifications:
   Polar, volatile
   Polar, non-volatile
   Non-polar, volatile
   Non-polar, non-volatile

Using the above criteria, two pesticides (lindane and diazinon), toluene, chloroform, gasoline and motor oil were used as representative model compounds for the validation experiments. In addition to being the most toxic substances available to the consumer, the two pesticides were also selected based on stability and detectability.

High levels of the intentionally contaminated bottles were used in the process to prepare REPETE polyester resin in order to verify the capability of the system to remove them. In every case, the amount of contaminated bottles added to the system was significantly higher than the estimated potential amount of contaminated bottles in the actual recycle stream.

Upon analysis of the final melt polymerized products prepared from the contaminated pcPET, no trace of the contaminants was found. This attests to the safety of the process. In all cases the test methods used were state of the art.

Physical and Chemical Properties of REPETE Polyester Resin. Table V shows physical properties of melt polymerized PET prepared using 10 and 20 weight percent pcPET compared to virgin PET prepared under the same conditions. These data suggest that the PET polymers containing up to 20% pcPET had properties equivalent to the control. The yellowness of the pcPET containing polymers is higher than the control and this is attributed to the additional thermal history experienced by the pcPET. It is also seen that the COOH end group concentration is lower for the polymers containing pcPET. This is most likely due to the use of excess EG during the pcPET glycolysis which would result in more hydroxyl end groups.

Table V. Physical Properties of Melt Polymerized PET Using pcPET

Property	Amount of pcPET		
	0%	10%	20%
IV (dl/g)	0.62	0.62	0.62
Color b	-0.8	2.6	4.1
Tm (°C)	254	253	253
COOH (eq/10^6 g)	41.1	32.0	25.9

It was also found that the melt polymerized polymers could be solid-state polymerized to higher molecular weight using typical reaction conditions for PET. The physical and chemical properties of the final polymers were investigated and the results shown in Table VI. The use of the pcPET resulted in an increase in the amount of DEG as well as in the various metals which were shown to be present in the pcPET used. If this data is compared with that for "as is" pcPET (see Tables II-IV) it is seen that the levels of catalysts and additives are raised only slightly by comparison. This is accomplished by adding catalyst based only on the virgin starting materials used in the process.

The data in Tables V and VI show the properties for the first generation of REPETE polyester resin. Subsequent to this, process optimization has led to an improvement in properties, especially in Gardner b color (yellowness). Polymer with b color in the range of -0.6 to 0.4 is now routinely obtained at the 25% pcPET level. These values are well within the commercial specification for Goodyear's virgin PET for the analogous product.

Table VI. Physical Properties and Chemical Composition Of Solid State Polymerized PET Containing pcPET

Property	Amount of pcPET		
	0%	10%	20%
IV(dl/g)	0.84	0.88	0.87
Color b	1.7	2.2	1.8
Tm (°C)	254	253	253
DEG (mole%)	1.22	1.39	1.63
Metals(ppm)			
Sb	208	196	222
Co	18	23	33
Mn	ND	ND	7
P	10	16	35

ND = Not Detected

Conclusion

Virgin PET has been modified with up to 20 weight % pcPET using a glycolysis/repolymerization process to produce a polymer with no significant changes in physical properties compared to 100% virgin PET. The chemical constituents of the pcPET are uniformly distributed and incorporated into the PET backbone. The additional thermal history experienced by the pcPET gave rise to a slight increase in the color in the first generation product. This technology is used to prepare REPETE polyester resin which is approved for use in both non-food and food container applications.

Acknowledgments

The authors acknowledge the valuable contributions of Gene Burkett, Randy-Sue Jenks, Ed Nowak, and Jim Ranomer.

References

1. Modern Plastics, Jan 1990, p 102.
2. Mangan, N.A., SPE Antec Proceedings, 27, 495 (1981).
3. Chemical & Engineering News, Jan 5, 1981, p 30.
4. Modern Plastics, April 1980, p 82.
5. Eastman Kodak Company, US Patent 5,051,528, (1991)
6. Mitsubishi Rayon, Japan, Patent 300314, 1990
7. Baliga, S. and Wong, W.T, J. Polym. Sci., Chem. Ed., 27, 2071 (1989).
8. Ostroski, H.S U.S. Patent 3,884,850, (1975).
9. Barham, V.F., J. Packaging Tech., Jan/Feb, 28, (1991).
10. Plastics Tech., April 1991, p 128.
11. Mitsui Mining & Smelting Co., Ltd., US Patent 4,132,633, (1979).

RECEIVED March 13, 1992

BLENDS AND APPLICATIONS

Chapter 16

Recycling Poly(ethylene terephthalate) in Unsaturated Polyester Composites

K. S. Rebeiz[1], D. W. Fowler[2], and D. R. Paul[3]

[1]Department of Civil Engineering, Lafayette College, Easton, PA 18042–1175
[2]Department of Civil Engineering, and [3]Department of Chemical Engineering, University of Texas, Austin, TX 78712

Recycled poly(ethylene terephthalate), PET, mainly recovered from used plastic beverage bottles, can be used to produce unsaturated polyester resins. Polymer concrete (PC) and polymer mortar (PM) made with these resins have been investigated for their properties, behaviors, and potential use in structural and engineering applications. Resins based on recycled PET offer the possibility of a lower source cost of materials for forming useful PC and PM based-products. The recycling of PET in PC and PM would also help solve some of the solid waste problems posed by plastics and save energy.

The condition of a nation's infrastructure is critically important for its economic vitality and international competitiveness. In a recent report, the National Council on Public Works Improvement identified solid waste, deteriorating highways, and disposal of hazardous wastes as three areas in the U.S. infrastructure of greatest needs for improvement *(1)*.

The development of new composite materials that are stronger, more durable, and more efficient to transport and apply than conventional materials is an essential element in the repair and improvement of the infrastructure. Polymer concrete (PC) and polymer mortar (PM) are examples of such new composite materials. PC and PM are formed from a resin and inorganic aggregates. PC uses coarse and fine aggregates while PM only uses fine aggregates.

Compared to cement-based materials, PC and PM are very strong and durable materials. The fast curing time of these products is another important advantage in many construction applications (PC and PM cure in a few minutes or hours while cement-based materials cure in a few days or weeks). However, the high cost of resins used in the production of PC and PM makes them expensive relative to cement-based materials and is the main disadvantage of the materials.

Unsaturated Polyester Resins Based on Recycled PET

Recently, some work has been done on the production of unsaturated polyester resins based on recycled poly(ethylene terephthalate), PET *(2)*. The PET wastes are typically found in beverage bottles that are collected after use in many U.S. areas. PET is depolymerized using ethylene, propylene, or neopentyl glycols. The resulting

0097–6156/92/0513–0206$06.00/0

glycolized products consist of oligomers and monomers. These oligomers are then reacted with unsaturated dibasic acids or anhydrides to form unsaturated polyester resins. A variety of proprietary ingredients may be added to give the resin some specific properties like flexibility or rigidity. Unsaturated polyester resins made from virgin materials usually use phthalic anhydride or isophthalic acid instead of terephthalate acid.

Unsaturated polyesters based on recycled PET might be a potentially lower source cost of resins for producing useful PC and PM based-products. The recycling of PET in PC and PM applications also helps in saving energy and providing for a long term disposal (more than 50 years) of the PET waste, an important consideration in recycling applications. This paper will therefore report on investigations of the important properties and behaviors of PC and PM using resins based on recycled PET.

Materials

Resins. Unsaturated polyesters based on recycled PET were supplied from several commercial sources. Some of the resins used were dark in color since the recycled PET used in their production was not purified to the extent other PET recycling applications might require. In most PC or PM applications, the purity of the recycled PET is not very important, which should minimize the cost of resins based on recycled PET.

Different formulations and different percentages of recycled PET were used in making the resins. The percentages of recycled PET used in these resins represent the minimum and maximum obtainable from the suppliers, which are 15 and 40 percent, respectively, of the alkyd portion of the resin (i.e., before styrene addition). The maximum percentage of recycled PET is usually desirable in making unsaturated polyesters because the PET residues do not adversely affect the polymer composite properties and their use helps reduce the cost of these products *(3, 4)*.

The resins, being prepolymers with high viscosity, were diluted with styrene. Styrene reduced resins viscosity to make them workable and allowed their further cure to a solid (polymer) by participating in the crosslinking via chain-reaction (addition) polymerization upon addition of free-radical initiators and promoters. As much as 50 percent styrene may be used in unsaturated polyesters without adversely affecting the physical and mechanical properties of the hardened materials *(5)*.

The use of resins with low viscosity is important because it results in resins with excellent wetting properties and the production of very workable polymer composites with high aggregate-to-resin ratio. A high aggregate-to-resin ratio is not only desirable for economy (since less resin, the expensive component, is used), but it also improves the mechanical and durability properties, the dimensional stability, and the heat resistance of the polymer composites.

Fillers. The properties of the polymer composites were strongly influenced by the type, properties, and amount of fillers. The amount of fillers varied from 10 to 90 percent by weight depending on the resin and filler properties, the intended use of the polymer composite, and how it is to be processed. A too small amount of filler disrupted the material homogeneity and hence decreased the strength of the composite. However, an increase in the amount of filler beyond a certain level increased the strength of the composite to a value exceeding the strength of the clear cast resin (high-filler effect). The maximum amount of filler was approximately 90 percent. Beyond this maximum value, the resin content was not enough to coat the fillers and the mechanical and physical properties of the polymer composite were sharply lowered.

The following coarse and fine inorganic aggregates were used in the experimental study of polymer composites based on recycled PET: 10-mm pea gravel, Colorado river siliceous sand with a fineness modulus of 3.26, and type F fly ash. The gravel and sand fillers were oven-dried for a minimum of 24 hours at 125 °C to reduce their moisture content to less than 0.5 percent by weight, thus ensuring good bond between the polymer matrix and the inorganic aggregates *(6)*. Fly ash, however, was already obtained dry from the supplier and therefore did not need to be oven-dried.

The use of fly ash, a waste material generated by the burning of coal, greatly improved the workability of the fresh mix. The fine and spherical particles of fly ash provided the fresh mix with better lubricating properties, thus improving its plasticity and cohesiveness. The better gradation obtained with fly ash also resulted in a hardened material with improved strength and mechanical properties.

Mix Design Optimization

The mix design for the clear cast resin was 1 percent (by weight of resin) of methyl ethyl ketone peroxide (MEKP) initiator and 0.1 percent (by weight of resin) of a 12 percent solution cobalt naphthenate (CoNp) promoter. Initiators and promoters were added to the resin immediately prior to casting.

The mix design for the PC and PM was optimized for workability, strength, and economy. PC mix was 10 percent resin, 45 percent 10-mm oven-dried pea gravel, 32 percent oven-dried sand, and 13 percent fly ash. PM mix consisted of 20 percent resin, 60 percent oven-dried sand, and 20 percent fly ash. One percent (by weight of resin) of MEKP initiator and 0.1 percent (by weight of resin) of 12 percent solution CoNp promoter were added to the resin immediately prior to mixing. Both PC and PM mixing was done using a conventional concrete mixer for a period of about three minutes. Specimens were then cast in molds and allowed to cure at room temperature. The age at testing of the specimens was three days unless otherwise specified.

Proper care should be taken during mixing to avoid direct contact between CoNp promoters and MEKP initiators because this mixture can react explosively. CoNp promoters should first be mixed thoroughly in the resin to ensure good dispersion. It is then and only then that the MEKP initiators can be safely added to the resin.

Testing

The clear cast resins (neat resins without the use of aggregates) were evaluated for their tensile properties according to ASTM D638 procedures. There are no standard tests that are directly applicable to PC and PM specimens. Therefore, in the evaluation of PC and PM mechanical and durability properties, ASTM standards applicable to cement-based materials were used as guidelines.

Compression specimens were 76-mm x 152-mm cylinders. Electrical strain gages, bonded to the specimens and connected to a data acquisition system, were used to read strains. The compression cylinders were tested in a hydraulic load machine at a constant rate of 44500 N/min. Flexure specimens were 50-mm x 50-mm x 305-mm beams. The beams were tested in third-point loading at a uniform rate of 2225 N/min. A dial gage was used to read the mid-span deflection of each beam.

Bond strength between PC or PM overlays and portland cement concrete substrate was measured using the pull-out test method *(7)*. Specimens were thin overlays (about 12-mm thick) cast directly on sandblasted portland cement concrete slabs. Circular grooves (100-mm diameter) were cored through the overlays and into the portland cement concrete substrate. Circular steel disks were then bonded to the sandblasted overlay at the cored locations using a strong epoxy. The disks were then pulled out in direct tension to determine the magnitude of the bond failure. The mix proportioning for the portland cement concrete substrate was designed to achieve a compressive strength of about 35 MPa with the use of air entraining agents.

The thermal expansion test used 76-mm x 152-mm cylinders. Longitudinal electrical strain gages were bonded to the specimens using a special epoxy system insensitive to high temperatures. The strain gages were then connected to a switch and balance unit in a full-bridge configuration. A piece of fused quartz with a known coefficient of thermal expansion was used as the compensating arm of the full-bridge circuit. The specimens were subjected to thermal cycles beginning at room temperature. The temperature was increased to 75 °C, decreased to -25 °C, and then returned to room temperature. Strain and temperature readings were taken in increments of 6 °C. The specimens were left at a constant temperature for a minimum of eight hours to ensure thermal stabilization before recording strains and temperatures.

Shrinkage testing did not follow ASTM guidelines applicable to cement-based materials because the shrinkage mechanism in PC and PM is different than in cement-based materials. PC and PM only experience short term shrinkage (plastic shrinkage) due to resin polymerization, whereas cement-based materials experience both short term shrinkage (plastic shrinkage) and long term shrinkage (drying shrinkage) due to water evaporation from the cement paste. A special device, shown in Figure 1, was therefore used to continuously monitor the shrinkage strains during the curing process. Specimens consisted of 76-mm x 76-mm x 305-mm beams cast inside Teflon-lined molds. The molds were wrapped in a plastic sheet to reduce the effect of ambient temperature changes on the plastic shrinkage readings. Immediately after mixing and placing the materials in the molds, the shrinkage measuring device was carefully inserted into the fresh PC or PM mix. The shrinkage device consisted of a horizontal rod to which two removable angles were attached. One angle was fixed while the other was free to move on roller bearings. A direct current differential transformer (DCDT), attached to the rod, was used to record the longitudinal displacement induced by shrinkage. PC and PM peak exotherm was continuously measured by means of thermocouples inserted inside the shrinkage specimens and connected to a digital temperature indicator.

Thermal cycling testing was done on thin PM overlays cast directly on portland cement concrete slabs. The specimens were put in an environmental chamber and subjected to thermal cycles. In each cycle, the temperature varied from -25 °C to 70 °C over a 24 hour period of time. The temperature was gradually increased and decreased to avoid thermal shock. Thermocouples, inserted inside the overlays and at different depths, were used to ensure that all specimens had the same temperature. At different thermal cycles, specimens were removed and the PM overlay was tested for its tensile bond strength to portland cement concrete using the pull-out test method.

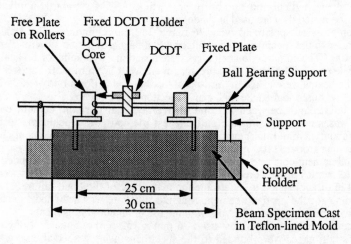

Figure 1. Shrinkage Testing

Results and Discussion

Fundamental Properties. Typical tensile properties for rigid and flexible clear cast resins are shown in Table I. These properties are very important in the potential use of these resins in PC and PM applications. Rigid resins, with high modulus and low elongation at break, are suitable in precast applications because they produce high modulus structures that deflect and creep less under the action of loads. Flexible resins, with low modulus and high elongation at break, are suitable in overlay applications because they provide compatibility to thermal and mechanical movements.

The typical properties of PC and PM using resins based on recycled PET are shown in Table II. These properties are comparable to those obtained from PC and PM using virgin resins and tested under the same conditions at the University of Texas.

PC and PM mechanical properties are different because PC and PM used a weight aggregate-to-resin ratio of 9-to-1 and 4-to-1, respectively. The difference in PC and PM properties enable these materials to be used in various applications where different properties are desirable. In precast applications, high strength, high modulus, and good dimensional stability are desirable. In overlay applications, low modulus and good bond strength to portland cement concrete are important properties.

Figure 2 shows the shrinkage strains induced by polymerization for PC and PM. The ultimate shrinkage strain of PM is about three times higher than the one corresponding to PC because more resin is available for polymerization in PM than in PC. Most of the shrinkage strains in PC and PM took place within the first eight hours after mixing and stopped after 24 hours. Shrinkage is important in many applications. In precast components, low shrinkage is important because excessive shrinkage strains may significantly affect the dimension of these structures, thus making their demolding, assembly, or use more difficult. In overlay applications, relatively low shrinkage is desirable because studies have reported that excessive shrinkage strains may cause delamination between the overlay and the substrate (8).

PC Time and Temperature Behavior. The time-temperature dependent properties of PC are important parameters to be accounted for in the design of precast structures. The early rate of strength gain is much faster for PC than it is for cement concrete. PC achieves more than 80 percent of its final strength in one day, while normal cement concrete achieves about 20 percent of its final strength in one day. Early strength gain is important in PC precast applications because these structures have to resist large stresses, early in their life, due to transportation and erection.

Loss in strength and stiffness result in the PC when subjected to high temperatures. An increase in temperature from 25 °C to 60 °C decreases the compressive strength by about 40 percent. PC is more susceptible to high temperatures than normal cement concrete because the synthetic viscoelastic resin binder used in producing PC is more temperature-sensitive than the inorganic cement binder used in producing normal cement concrete. However, despite this loss in strength, PC remains a strong material compared to normal cement concrete. For example, the flexural strength of PC at 60 °C is 14 MPa, while the flexural strength of normal cement concrete is about 6 MPa.

TABLE I. Typical Tensile Strength Properties of Clear Cast Resins

Tensile strength properties	Rigid resin	Flexible resin
Tensile strength (MPa)	21.1	13.3
Young's modulus (GPa)	1.24	0.30
Elongation at break (%)	2.4	53.5

TABLE II.   Typical Properties of PC and PM Specimens

Properties	PC	PM
Compressive strength (MPa)	91.8	72.2
Compressive modulus (GPa)	28.6	6.7
Flexural strength (MPa)	18.7	14.6
Flexural modulus (GPa)	26.6	4.4
Tensile bond strength to portland cement concrete (MPa)	1.9	2.6
Ductility index	2.8	4.5
Coefficient of thermal expansion for $20^{\circ}C \leq T \leq 70^{\circ}C$ (10^{-6} mm/mm/$^{\circ}$C)	14.9	41.7
Poisson's ratio	0.27	0.32
Shrinkage (10^{-3} mm/mm) at 24 hours	2.1	4.9
Peak exotherm ($^{\circ}$C)	31	40

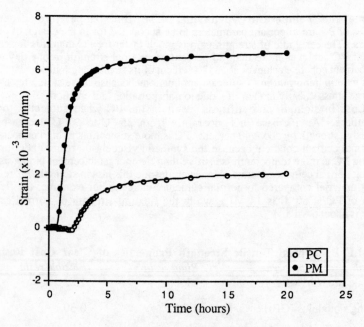

Figure 2. PC and PM Shrinkage

PM Thermal Cycling. Variations in temperature result in high internal stresses at the interface between the PM overlay and the portland cement concrete substrate. The internal stresses are developed because the coefficient of thermal expansion of the PM overlay is high compared to the one corresponding to the portland cement concrete substrate. The repeated changes in temperature, or thermal fatigue, may eventually reduce the bond strength between the overlay and the portland cement concrete substrate, thus causing delamination of the overlay and increasing its permeability to water and chloride ions. The thermal cycling test is therefore an accelerated test procedure designed to simulate the effect of large variations in temperature inherent to structures such as pavements and bridges.

Thermal cycling test results showed that PM using polyester resins which have an elongation at break of less than 10 percent do not perform as well as PM using polyesters with an elongation of 20 to 50 percent. Low modulus resins binders are desirable in PM overlays because they produce low modulus PM, thus offsetting the effect of the higher coefficient of thermal expansion *(9)*. Resins with high elongation at break, capable of stretching during thermal movements, are also desirable.

PC Applications. PC using resins based on recycled PET could be among the best systems available today for the disposal of hazardous wastes in precast structures. The fast cure time of the material enables the fast and efficient use of forms and other production facilities. The very good strength and durability properties permit the use of thinner sections, thus reducing dead loads in structures and minimizing transportation and erection costs. The very low permeability of the materials to liquids and gases (unlike portland cement concrete, a porous material) and its inherent corrosion resistance (unlike metallic materials) allow for the safe disposal of toxic and acid wastes over long periods of time.

In power line transmission poles, the high flexural strength of PC may eliminate the need to use reinforcing steel (if reinforcement is needed, glass pultrusion bars may be used instead of steel bars because of their electrical resistance). The absence of water in PC along with its low permeability contribute even further to its very good dielectric properties. In high voltage insulators, PC has been used as a replacement of porcelain to take advantage of the dielectric properties and the ability to produce very smooth surfaces.

Machine bases and machine tool components could also become very significant uses of PC. The ability to cast complex shapes which reduces the tooling cost, the excellent damping characteristics, the high stiffness-to-weight ratio, the low thermal conductivity, and the corrosion resistance makes PC a very competitive material relative to cast iron, the material traditionally used.

PM Applications. PM using resins based on recycled PET is a very good material in the repair of pavements and bridges. PM overlay provides low water and chloride permeabilities. Thus, it helps prevent deterioration and spalling of concrete due to freezing and thawing cycles and corrosion of the reinforcing steel. It also acts as a surfacing material to repair spalled and deteriorated concrete and maintains high skid resistance, thus improving ride quality and safety. The very good mechanical and durability properties allow the use of less materials, and, therefore, minimizes dead load on the structures, reduces clearance problems, eliminates the need to reconstruct approach slabs for bridges, and reduces maintenance and repair activities. The very fast cure time of the material permits overnight applications and the pavement or bridge can be returned to traffic the next morning. This property results in minimum traffic disruption, less delays, greater safety, and eliminates the need to construct expensive detours. Different methods are used to apply PM overlays in pavements and bridges. In one method, the resin is first applied over a clean and sound surface. Immediately after applying the resin, dry sand is applied and polymerization takes place. Excess sand is then brushed away. The process is repeated until three or four layers have been applied. Another method consists of premixing PM and applying it over the surface using paving machines.

PM coatings can be very successfully used to protect normal cement concrete or steel tanks, pipes, or containers from aggressive liquids or radiations from nuclear wastes. The overlay physically reinforces the structure and provides a very strong, wear-resistant, low permeability, and acid-resistant surface. It is lightweight, cures rapidly, and provides good bond strength to portland cement concrete. In very aggressive environments, specially formulated polyester resins are used in making PM to provide the necessary resistance against the action of heat and flame, strong chemicals, or other aggressive agents. The development of shotcrete equipments permit the easy use of thin PM coatings on vertical, horizontal, and irregular surfaces.

PM could also find applications as adhesives for joining different components of same or different composition. For example, PM adhesives could be used to bond new concrete to old concrete, or to bond old concrete to steel plates, steel bars, or heavy equipment machinery. Low modulus, high elongation resins should be used in joints where movement must be allowed.

Conclusions

PC and PM using resins based on recycled PET are technologically and economically attractive systems in a variety of engineering and construction applications, especially at a time when the infrastructure is in urgent need for repair. The formulation of resins based on recycled PET can relatively easily be altered to achieve a wide variety of properties and performances for the polymer composites. In most of the engineering applications, the purity of the recycled PET is not very important, a fact which should minimize the cost of resins based on recycled PET. The use of fly ash, a waste material produced by the burning of coal, was found to improve the workability of the fresh PC and PM mix and to increase their hardened strength properties. The main disadvantage of PC and PM is the possibility of loss in strength due to high temperatures, an important factor to be considered in certain applications such as precast building panels.

Acknowledgment

The authors acknowledge the support for this research from the Advanced Research Program of the Texas Higher Education Coordinating Board.

Literature Cited

1. "Fragile Foundations: A Report on America's Public Works," National Council on Public Works Improvement, Final Report to the President and Congress, **1988**.
2. Vaidya, U. R., and Nadkarni, V. M. "Unsaturated Polyester Resins from Poly(ethylene terephthalate) Waste," *Industrial & Engineering Chemistry Research,* **1987**, *Vol. 26* , pp 194-198.
3. Rebeiz, K. S., Iyer, V. S., Fowler, D. W., and Paul, D. R., "Polymer Concrete Using Resins With Recycled PET," Proceedings 48th Annual Technical Conference of the Society of Plastics Engineers (ANTEC 90), **1990**.
4. Schneider, J. B., Ehrig, R. J., Brownell, G. L., and Kosmack, D. A., "Unsaturated Polyesters Containing Recycled PET," Proceedings 48th Annual Technical Conference of the Society of Plastics Engineers (ANTEC 90), **1990**.
5. Ohama, Y., Demura, K., and Komiyama, M., "Properties of Polyester Resin Concrete with Various Styrene Contents, Proceedings of the Twenty-Second Japan Congress on Materials Research, The Society of Materials Science, Japan, **1979**, pp 360-363.

6. Nutt, W. O., and Staynes, B. W., "The Next Twenty Five Years," Proceedings 5th International Congress on Polymer Concrete, Brighton, England, **1987**, pp 413-416.
7. "Use of Epoxy Compounds with Concrete (ACI 503R-89)," American Concrete Institute (ACI) Committee 503, American Concrete Institute, Detroit, **1989**.
8. Al-Negheimish, A., "Bond Strength, Long Term Performance and Temperature Induced Stresses in Polymer Concrete-Portland Cement Concrete Composite Members," Ph.D. Dissertation, The University of Texas, Austin, Texas, **1988**.
9. Sprinkel, M. M., "Thermal Compatibility of the Polymer Concrete Overlays," *Transportation Research Record*, **1982**, *No. 899*, Highway Research Board, pp 64-73.

RECEIVED March 17, 1992

Chapter 17

High-Density Polyethylene—Poly(ethylene terephthalate) Blends

Compatibilization and Physical Properties

S. A. Jabarin, E. A. Lofgren, and S. B. Shah

Polymer Institute, College of Engineering, University of Toledo,
Toledo, OH 43606–3390

Blends of poly(ethylene terephthalate), PET, and high density polyethylene,HDPE, have been investigated with regard to their mechanical, rheological, melting, crystallization, and morphological characteristics.

Binary blends of HDPE and PET exhibit very poor processing and mechanical properties due to their complete incompatibility. With the addition of maleic anhydride olefin grafted copolymer as a compatibilizing agent, however, ternary blends can be obtained which display good processability, mechanical, and optical properties. The mechanical properties of these blends have been found to be specifically dependent upon composition as well as morphological phase characteristics. Variation of these factors can result in four different levels of impact and elongation to break properties.

The crystallization kinetics of the PET component in the ternary blends were studied under dynamic cooling and isothermal conditions. The role of the polyolefin components of the blends, in affecting the rate of crystallization of PET, is discussed with regard to composition and mechanism of crystallization.

Recent advances in recycling technology related to separation and cleaning have provided industry with clean useable recycled streams of poly(ethylene terephthalate), PET, and high density polyethylene, HDPE, materials. In certain applications, it is desirable to take advantage of PET's inherent chemical resistance and HDPE's processing characteristics as related to extrusion and extrusion blow molding. Blends of HDPE and PET may provide a successful route to improving the extrusion blow molding characteristics of PET. The blends, however, exhibit very poor mechanical properties due to the incompatibility of these two polymers.

0097–6156/92/0513–0215$06.00/0
© 1992 American Chemical Society

Previous work (1-6) has shown that polymer mixtures may be mechanically compatibilized by the addition of another polymer. Such an approach was applied (6) to blends of HDPE/PET with the addition of small amounts of commercially available triblock copolymers of styrene and butadiene. The butadiene rubber block provides a certain miscibility with the polyethylene backbone and the end blocks of styrene, which exhibit aromatic character similar to that of PET, causes the miscibility with PET (6,7).

The objectives of this study are to evaluate the utilization of other compatibilizing agents for improving the extrusion and extrusion blow molding characteristics of HDPE/PET blends and also to improve their physical and mechanical properties. The primary compatibilizing agent chosen for this study was maleic anhydride grafted polyolefin resin. The choice of this compatibilizer is based on the assumption that the polyolefin chain provides the miscibility with HDPE and the anhydride group provides the specific interaction with the hydroxyl end groups of PET.

Experimental

Materials. Various combinations of PET, HDPE and compatibilizing agents were blended and made into sheets suitable for further evaluation. Materials included in this study were:

PET:	Hoechst T-80	(0.8 intrinsic viscosity)
	Goodyear 8006	(0.8 intrinsic viscosity)

HDPE: Phillips EHM 6003

Compatibilizing agent: Maleic anhydride modified polyolefin of the kind (Mitsui Admer AT 4696) having a melt flow rate of 1.3 measured at 190°C and a density of 0.88 gm/cm^3

Blend Preparation. A Werner and Pfleiderer Corporation (ZK-30) self wiping co-rotating twin screw extruder was used to melt blend Goodyear PET with HDPE and compatibilizing agent. Before blending, the PET resin was air dried at 150°C in a Bry Air Company (D-75) hopper dryer system. After 8 hours, the temperature was reduced to 60°C, the lower melting resins were added, and drying was continued for an additional two hours. The twin screw (diameter 30mm) extruder utilized a twin strand die and was operated at 300 rpm and 100 psi. Melt zones 1 and 2 were held at 230 and 250°C respectively, while zones 3-5 were held at 270°C. Extrudate was passed through cooling water and pelletized.

Sheet Preparation. Pelletized melt blended material plus any additional compatibilizing agents were vacuum dried overnight at 105°C, before being prepared as narrow sheet or ribbon using a Brabender Laboratory scale single screw extruder. A general purpose screw (diameter 19mm) was used at 50 rpm with temperatures of 240°C for zone 1 and 270°C for zones 2, 3 and the die. A take-off speed of 15 rpm was used to prepare ribbon samples 31.8 mm wide and 0.38-0.51 mm thick.

Hoechst PET was dry blended with various combinations of HDPE and compatibilizing agent before being vacuum dried at 55°C for 48 hours. Each blend was then extruded with the Brabender at 70 rpm with zone 1 at 260°C, zones 2 and 3 at 280°C and the die at 250°C. A take-off winder speed of 15 rpm was used to prepare ribbon samples 31.8 mm wide. Unblendend and ternary blend ribbons were 0.38-0.51 mm thick, while binary blend ribbons had thickness up to 1.09 mm.

Thermal Properties. Thermal properties of vacuum dried, blended sheet materials were monitored under dynamic and isothermal conditions using a Perkin-Elmer Differential Scanning Calorimeter (DSC-2). Analyses were performed under a nitrogen purge to prevent oxidative degradation. Dynamic behavior was monitored at heating or cooling rates of 10°C per minute. Isothermal crystallization was monitored by quickly cooling to the desired temperature after holding the sample at 300°C for 10 minutes, to remove all previous crystallinity and thermal history. Melting and crystallization transitions for HDPE and PET were measured individually. Area calculations were performed using standard Perkin-Elmer TADS 3600 computer software and reported as heat of fusion or crystallization.

Physical Properties. Tensile mechanical properties were determined at 23°C using a table model Instron 1101 Tensile Tester according to a modified microtensile ASTM Procedure 1708. Crosshead speeds of 25.4 mm/min. were used to pull die cut samples 4.75 mm wide, with thicknesses from 0.38-1.09 mm and gauge lengths of 22.3 mm.

Tensile impact energy was measured at 23°C using a swinging pendulum type, Custom Scientific Tensile Specimen-in-Base Impact Tester. ASTM Procedure 1822 was followed using previously described die cut specimens.

Rheological Properties. Melt viscosity and shear sensitivity were monitored at 280°C under a nitrogen atmosphere using a Rheometrics Visco-Elastic Tester (RVE) equipped with a 25 mm cone and plate sample holder. Shear rates from 0.1 to 100 radians per second were utilized.

Microscopy. Morphological structures of blend samples were examined using microscopy techniques. An Hitachi S-2700 Scanning Electron Microscope (SEM) was used to observe surface characteristics of samples freeze fractured in liquid nitrogen. Fractured samples were sputter coated with 8 um of gold-palladium before analyses.

Results and Discussion

Preparation of Blends. Blends were prepared using a Twin-Screw Extruder. For the binary blends of HDPE/PET, without a compatibilizing agent, pellets were obtained from the extrudate strands. However, when extruded using a Brabender extruder equipped with a ribbon die, at conditions used to extrude ternary blends and unblended materials, the extruded ribbons were not acceptable because of their very weak mechanical properties and their physical appearance. The extruded binary blend samples exhibited surfaces full of holes indicating complete incompatibility. On the other hand, when using a maleic anhydride grafted polyolefin as a compatibilizer, good extruded ribbons were obtained, which exhibited both good mechanical and surface properties. Several binary blend samples were prepared by adjusting processing conditions, reducing "take up" roll speed, and increasing ribbon thickness. These samples were easily delaminated and appear d to be composed of many distinct layers. Compositions of prepared blends are shown in Table I.

Table I. Blend Compositions (in percent weight) of Ternary and Binary Blends of PET/HDPE/Compatibilizing Agent

Sample No.	Percent Weight		
	PET	HDPE	Maleic Anhydride Grafted Olefin
1	100	-	-
2	85	10	5
3	80	15	5
4	75	20	5
5	65	30	5
6	55	40	5
7	45	50	5
8	35	60	5
9	25	70	5
10	15	80	5
11	5	90	5
12	-	100	-
13	85	15	-
14	50	50	-
15	15	85	-

Mechanical Properties. The tensile mechanical properties of the various blends and the two pure materials of PET and HDPE are given in Figures 1-3 for elongation to break, yield stress and modulus respectively. Figure 1 gives the percent elongation to break as a function of PET composition in ternary and binary blends. For ternary blend samples, it is seen that this property is a complex function of blend composition, exhibiting an apparent minima at 75% PET. Although the minima occurs at this composition, the percent elongation to break is about 50% which is still much higher than the values obtained for blends without a compatibilizing agent. It is also worth noting that the elongation to break of the blends containing 85% PET, 10% HDPE and 5% maleic anhydride indicates mechanical properties only slightly less than those of pure PET. The properties of the ternary blends at all compositions are quite good, compared to the properties of the binary blends without compatibilizing agent shown as the dashed line. These results show the importance of the compatibilizing agent and they indicate that the maleic anhydride grafted olefin copolymer is providing a limited miscibility with other components to cause

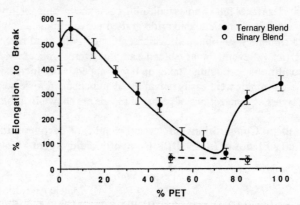

Figure 1   Percent elongation to break of ternary and binary blend samples as a function of PET concentration

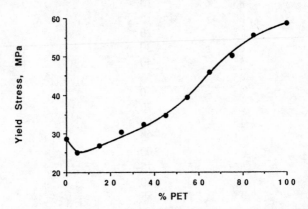

Figure 2   Yield stress of ternary blend samples as a function of PET concentration

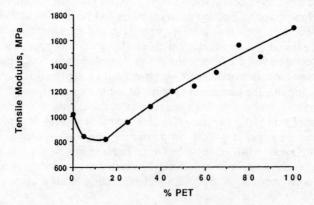

Figure 3   Tensile modulus of ternary blend samples as a function of PET concentration

improvements in mechanical properties. The behaviors of the yield stress and modulus, as a function of blend composition, are given in Figures 2 and 3. Both of these properties are shown to decrease with decreasing amounts of PET or increasing amounts of HDPE in the ternary blends with an apparent minima at the 15% PET or 80% HDPE composition.

The impact strength as a function of blend composition is given in Figure 4. The impact properties of the ternary blends are good compared to the properties of the binary blends without compatibilizing agent. The binary blends exhibit very poor impact properties at all compositions.

Rheological Properties. In order to process a material by free extrusion blow molding, that material must have sufficient melt strength. Melt strength is a material property which describes the ability of a free hanging parison to retain its shape. A low melt strength material will experience parison sag, making it impossible to blow a good container. It is therefore necessary that the melt strength be as high as possible.

Currently, processing PET by extrusion blow molding is limited by a lack of melt strength. PET will "drool" into a pool of material. Raising the melt strength of PET involves changing the rheological characteristics of the material, which can be achieved by blend modification. The PET blends, however, must exhibit greater melt strength and also must be strong under end-use conditions.

Melt strength is in essence the extensional viscosity, which is in turn proportional to the zero-shear viscosity. Because it is not possible to measure the zero-shear viscosity directly, a low shear rate of 0.1 rad/s is chosen as an indication of melt strength.

Viscosity is also important for transport of polymer melt through the flow channels in the blow molding machine. When the viscosity is high, at high shear rates, such as those encountered during processing, the pressure, torque, and temperature of the melt becomes excessive. It is therefore desirable to have low, high-shear viscosity.

Shear sensitivity, the ratio of low-shear to high-shear viscosity, describes the overall processibility of a resin. A highly shear sensitive material, which is desirable, will have a good balance of high melt strength and low extrusion pressure.

Figure 5 gives the viscosity as a function of shear rate for pure PET, HDPE and ternary blends of PET/HDPE/compatibilizing agent. It is seen that pure PET exhibits a low viscosity at low shear rates as well as a very low shear sensitivity compared to pure HDPE resin. The ternary blends show significant improvements in both low shear viscosity (melt strength) and shear sensitivity. A summary of the values for the viscosities at low and high shear rates as well as the values of the shear sensitivities of the various compositions studied are given in Table II. These results indicate that adding HDPE with compatibilizing agent to PET, causes great increases in the low shear viscosity or melt strength of PET. In addition, the shear sensitivity or processibility of the ternary blends increases with increasing amounts of HDPE in the blends.

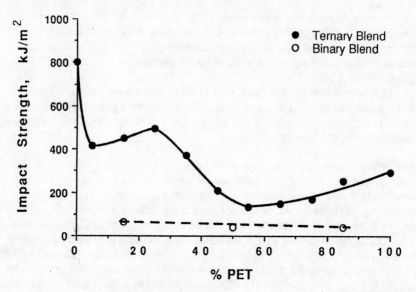

Figure 4   Impact strength of ternary and binary blend samples as a function of PET concentration

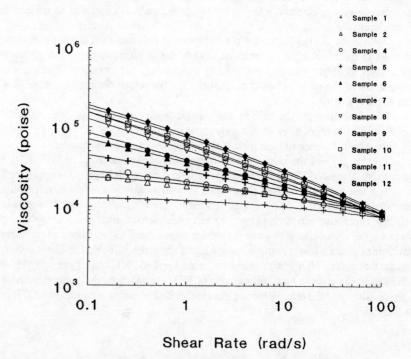

Figure 5   Melt viscosity of various ternary blend samples as a function of shear rate

**Table II. A Summary of the Viscosities at Shear Rates of
0.1 and 100 rad/sec. as well as Shear Sensitivity for the
Various Blend Compositions Measured at 280°C
(For Sample Composition see Table I)**

Sample No.	Viscosity 0.1 rad/s (Kpoise)	Viscosity 100 rad/s (Kpoise)	Shear Sensitivity (1/100)
1	13.6	7.2	1.6
12	184.2	8.9	9.0
2	24.8	8.3	2.2
4	31.5	7.4	2.8
5	46.0	8.2	3.4
6	72.6	8.4	4.3
7	85.5	8.1	4.7
8	122.7	7.6	6.7
9	148.8	8.1	7.5
10	152.9	8.1	7.8

Thermal Properties. Melting and crystallization characteristics of the blend samples were monitored using Differential Scanning Calorimetry (DSC). Heating scans of these materials exhibit independent melting and crystallization transitions, which are characteristic of the pure components of PET and HDPE. The thermal behavior of the compatibilizing agent, which is completely melted at temperatures above 133°C, is masked by that of the HDPE component. Figure 6 shows melting peak temperatures obtained for PET and HDPE components of the ternary blend samples. These results were obtained while heating samples, previously crystallized from the melt at a cooling rate of 10°C per minute. At all blend compositions, peak melting temperatures of both components are seen to remain separate and independent of concentration, with no suppression of melting temperatures.

Heats of crystallization (ΔH_c) were recorded while cooling blend samples from the melt at 10°C per minute. Results shown in Figure 7 indicate that heats of crystallization, measured for the PET component of ternary blends, are a function of blend composition. As the proportion of PET in the ternary blend is reduced, the ΔH_c of the PET (normalized to per Kg PET in the blend) is also reduced. At concentrations of less than 35% PET (60% HDPE and 5% compatibilizing agent), ΔH_c values obtained for PET approach zero. This behavior would not be expected because if two polymers are immiscible in the melt, the non-crystallized polymer diluent (HDPE) does not affect the crystallization of the other polymer (8). Binary blends, prepared without compatibilizing agent, do not show reduced ΔH_c values as concentrations of PET are reduced. These results, therefore, suggest that the compatibilizing agent is providing a limited amount of miscibility between the components of the blends. Similar reduction in degree of crystallinity with decreasing concentration of the crystallizable component were observed by Ong and

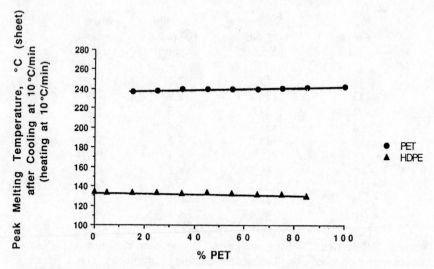

Figure 6   Peak melting temperatures of PET and HDPE as a function of PET concentration

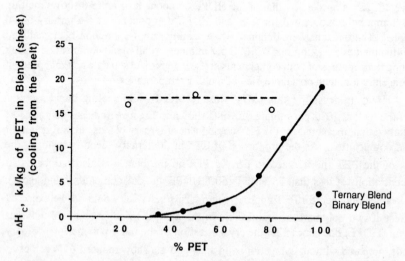

Figure 7   Magnitude of crystallization exotherms of PET component in ternary and binary blends as a function of PET concentration

Price (8,9) and further described by Stein et al (10) for binary blends of crystallizable poly(ε-caprolactone), PCL, and non-crystallizable poly(vinyl chloride), PVC. They concluded that the amorphous PVC must be included within the PCL spherulites, causing these changes in crystallization behavior. Results shown in Figure 7 may be explained in the same manner for the ternary blends of PET, HDPE and the compatibilizing agent. Part of the polyolefin component may penetrate the PET crystal structure, causing the observed reduction in crystallization of PET as the concentration of HDPE is increased.

An additional means of monitoring crystallization behavior is by noting changes in peak crystallization temperature, which is related to crystallization rate. Figure 8 gives changes in PET peak crystallization temperature as a function of PET concentration in the ternary blends. As with ΔH_c values, peak temperatures are shifted to lower temperatures with decreased concentrations of PET. This indicates reduced rates of crystallization and is consistent with results reported by Ong and Price (8,9). Crystallization temperature remains constant for binary blends of PET and HDPE, without compatibilizing agent.

Heats of crystallization and peak crystallization temperatures recorded for the HDPE component of the blends remain relatively constant over the entire range of concentrations.

Ternary blends were also crystallized from the melt while being held isothermally at 180 and 190°C. These conditions permit the PET portion of each blend to crystallize thoroughly, while the HDPE and the compatibilizing agent remain in the melt phase. Isothermal peak crystallization times can be taken as a measure of crystallization half-times. Figure 9 shows results obtained at 180°C and represents trends observed at both isothermal crystallization temperatures. Peak crystallization times, obtained for the PET portion of the blend samples, are found to change as a function of composition. This behavior, observed during isothermal crystallization, is different from that observed for samples crystallized under conditions of dynamic cooling from the melt. It appears that two mechanisms may be operating to control isothermal crystallization from the melt. At low levels of HDPE in the ternary blend (10-20%), the crystallization rate of PET is reduced from that of unblended PET. At higher concentrations of HDPE, (30-60%), the rate of crystallization of PET increases. Stein, et al (10) observed a similar half-time maximum or minimum in crystallization rate, for blends of poly(butylene terephthalate) and poly(ethylene terephthalate), using a depolarized light intensity technique. Ong and Price (8) have reported that when PVC is blended with PCL, half-time of PCL crystallization increase with increased concentration of the amorphous PVC phase. These results are similar to those observed for our ternary blends at low concentration of molten HDPE phase and high levels of crystallizable PET phase.

The ternary blend results indicate that at low concentrations of HDPE, this component is interfering with or penetrating the spherulitic structure of PET, and thus reducing its crystallization rate. At higher concentrations of HDPE, molten polyethylene composes a greater portion of the matrix, while the PET component is reduced. This increased portion of molten HDPE appears to provide a greater mobility for the PET, allowing it to crystallize more readily. Further work is required

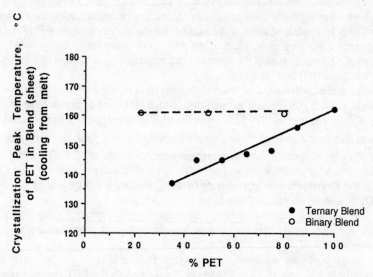

Figure 8   Crystallization peak temperature of PET component in ternary and binary blends as a function of PET concentration

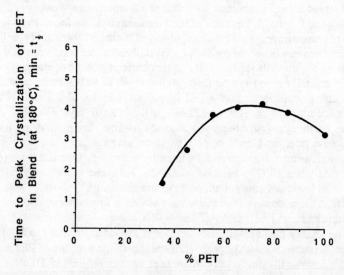

Figure 9   Time to isothermal peak crystallization at 180°C for PET in ternary blend samples as a function of PET concentration

to follow spherulitic growth and development during isothermal crystallization as well as to employ wide angle x-ray diffraction measurements to elucidate the nature of the crystallization process.

Morphology. Morphological features of the ternary blends were examined using scanning electron microscopy. Typical fracture surfaces of extruded sheets (ribbons) are shown in Figure 10 for the unblended PET and HDPE. No specific morphology can be detected. Figure 11 gives photomicrographs of ternary blends containing various amounts of PET and HDPE. In all four cases highly oriented features of gross microstructures are observed. For sample "a", with 85% PET (blend no. 2), the oriented phases appear to be homogeneously dispersed and interconnected.

As the ternary blend composition of PET decreases, and that of HDPE increases, the oriented characteristics of the structures are maintained. There are indications, however, that within the oriented phase morphology, individual segregates of oriented phase morphologies are obtained. Examples of these phenomena are seen with samples "b-d" of Figure 11 (PET compositions from 75-35%). Across the thickness of these samples one region is seen to exhibit a smooth texture while another shows a more fibrillar type morphology. Morphological features of this nature have also been observed by Trangott et al (6), and described as "phase orientation". They found this "phase orientation" to be "distinct from molecular orientation within a phase" and to result from "flow during processing".

These morphological features can be well correlated with the mechanical properties discussed above. Specifically it was concluded that in terms of impact strength and elongation to break, blend no. 2 ("a" or 85% PET), shows excellent mechanical properties. Blends no. 4 and 5 ("b" and "c" or 75 and 65% PET) exhibit lower mechanical properties than blend no. 2. With further reduction of PET concentration and subsequent increase in that of HDPE, mechanical properties are noted to increase as indicated with blend no. 8 ("d" or 35% PET/60% HDPE). In this case, the higher concentration of HDPE overrides the reduction in mechanical properties resulting from phase segregation, due to the higher impact strength and elongation properties of HDPE. These results indicate therefore, that mechanical properties are affected by the compositions of the blends as well as by their oriented phase morphology.

During this study of the evaluation of ternary blends' mechanical properties it was observed that certain blend compositions sometimes exhibited mechanical properties lower than those recorded for any of the blends presented on Table I and discussed with Figure 11. Upon further examination, photomicrographs of such samples revealed extremely interesting types of morphological features. Figure 12 gives a comparison of the morphology of the samples exhibiting various mechanical properties. Sample "a" of this figure shows blend no. 3 with highly oriented homogeneous phase morphology. This ternary blend sample gives an elongation to break of about 200%. Samples "b", "c" and "d" were prepared with the same composition as sample "a", namely 80% PET, 15% HDPE, and 5% compatibilizing agent. The elongation to break of these samples is only about 10%. The morphological character of samples "b", "c" and "d" are indicative of unoriented phase separated morphology, even though the compatibilizing agent is providing

100% PET

100% HDPE

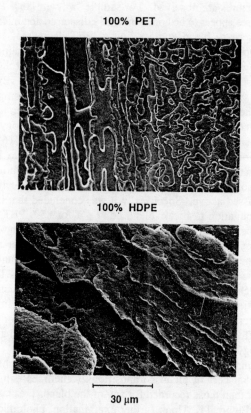

30 μm

Figure 10   Scanning electron micrographs of fractured surfaces of PET and HDPE

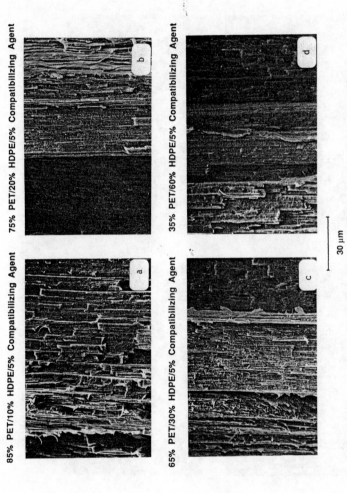

Figure 11   Scanning electron micrographs of fractured surfaces of ternary blends containing various amounts of PET and HDPE

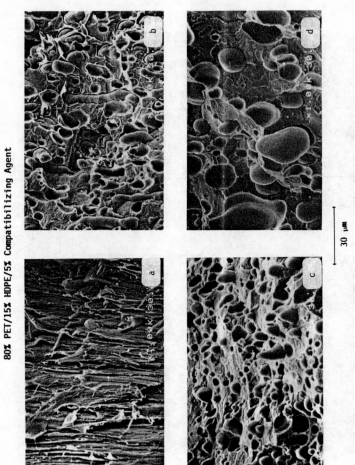

Figure 12   Scanning electron micrographs of fractured surfaces of ternary blends of various oriented morphologies

some binding between the phases. In these micrographs the matrix is PET and the semispherical structures are HDPE. These results indicate that the mechanical properties of the ternary blends are controlled by the amounts of orientation present, as well as by blend composition.

Summary and Conclusions

The compatibilization and properties of blends prepared from PET and HDPE have currently been investigated as functions of: compatibilizing agent, blend composition, and processing conditions. Binary blends of PET and HDPE exhibit extremely poor mechanical and optical properties indicating that they are totally incompatible. Because of these properties, as well as processing difficulties, such binary blends serve no useful purpose. A maleic anhydride olefin grafted copolymer is a third component which may be added to blends of PET and HDPE. The addition of this compatibilizing agent yields ternary blends which can be processed to form useful products.

Ternary blends compounded with various ratios of PET and HDPE exhibit good mechanical properties and good melt strength for uniform extrusion and extrusion blow molding processes. Mechanical properties of these blends have been found to be specifically dependent upon composition as well as morphological phase characteristics. Variations of these factors can result in four different levels of impact and elongation to break properties.

a) High impact and elongation to break values can be achieved with ternary blends composed of oriented, homogeneously dispersed, interconnected phases, created with appropriate processing conditions.

b) Lower mechanical properties, which are still greatly superior to those of binary blends, are illustrated with ternary blends containing oriented but somewhat segregated phases.

c) Increased mechanical properties are attained, in the presence of segregated phase orientation, as ternary blend compositions of HDPE are increased.

d) Reduced mechanical properties are exhibited by ternary blends in the absence of orientation, although the phases are interconnected with compatibilizing agent.

Thermal property evaluations of ternary blends reveal the influence of compatibilizing agent, during crystallization in both dynamic and isothermal modes. These results show the compatibilizing agent provides the miscibility required to improve mechanical properties of blends prepared from PET and HDPE.

Dynamic crystallization data recorded for ternary blends indicate amount of crystallinity achieved for the PET component of blends, decreases as blend concentration of HDPE increases. This behavior indicates that the polyolefin component has penetrated the spherulitic structure of PET.

Isothermal crystallization results illustrate additional effects of the polyolefin component on PET crystallization behavior. Under these conditions two mechanisms are seen to occur. At low blend levels of HDPE, rate of PET crystallization is reduced because of interference of the polyolefin in the crystalline structure of PET.

At higher levels of HDPE in the blend, with the HDPE component becoming the major portion of the matrix, rate of PET crystallization increases. In this case, the HDPE matrix appears to provide greater mobility for the PET, allowing it to crystallize more readily.

Literature Cited

1. Paul, D.R., In *Polymer Blends*, Paul, D.R. and Newman, S. Eds.; Academic, New York; 1978, Vol. II, Chap. 12.
2. Nolley, E.; Barlow, J.W.; and Paul, D.R. *Polym. Eng. Sci.*, **1980**, 20, 364.
3. Lindsey, C.R.; Barlow, J.W.; and Paul, D.R. *J. Appl. Polym. Sci.*,**1981**, 26, 9.
4. Lindsey, C.R.; Paul, D.R.; and Barlow, J.W. *J. Appl. Polym. Sci.*, **1981**, 26, 1.
5. Bartlett, D.W.; Paul, D.R. and Barlow, J.W. *Mod. Plast.*, **1981**, 58 (12), 60.
6. Traugott, T.D.; Barlow, J.W. and Paul, D.R. *J. Appl. Polym. Sci.*, **1983**, 28, 2947.
7. Barlow, J.W.; and Paul, D.R. *Polym. Eng. Sci.*, **1984**, 24, 525.
8. Ong, C.J. and Price, F.P. *J. Polym. Sci.*, Symposium, **1978**, 63, 59.
9. Ong, C.J. and Price, F.P., *J. of Polym. Sci.*, Symposium, **1978**, 63, 45.
10. Stein, R.S.; Khambatta, F.B.; Warner, F.P.; Russell, T.; Escala, A. and Balizer, E. *J. Polym. Sci.* Symposium, **1978**, 63, 313.

RECEIVED April 28, 1992

Chapter 18

Composite Materials from Recycled Multilayer Polypropylene Bottles and Wood Fibers

Rodney J. Simpson[1] and Susan E. Selke

School of Packaging, Michigan State University,
East Lansing, MI 48824-1223

The feasibility of combining recycled multi-layer polypropylene (PP) bottles with untreated hardwood aspen fiber was tested by evaluating mechanical properties of reclaimed polymer and virgin PP composites. The bottles consisted of 3.75% ethylene vinyl alcohol, 1.75% adhesive, and 94.5% random copolymer ethylene and propylene. The wood fiber was obtained from thermomechanical pulp (TMP). Up to fifty weight-percent of wood fiber was incorporated into the matrix, utilizing twin-screw extrusion followed by compression molding. Orientation of the wood fiber improved mechanical properties. The PP-reclaim-wood fiber composite was superior to the virgin PP composite, possibly due to increased adhesion at the interface. The multi-layer material also exhibited better dimensional stability under extreme environmental conditions.

As solid waste disposal problems become more acute, there is increased interest in recycling as an alternative to disposal. Plastic packaging is highly visible as a waste management problem due to its overall volume percent and short life span. Increasingly, plastic packaging is being included in collection programs for recycling, resulting in a rapid increase in availability of recycled thermoplastics. Appropriate markets for utilization of recovered materials remains a major concern. Multilayer plastic bottles are seen as particularly problematic, since the combination of resins can lead to significant deterioration in properties in the recovered materials.

[1]Current address: Colgate-Palmolive Company, Piscataway, NJ 08855

0097–6156/92/0513–0232$06.00/0
© 1992 American Chemical Society

Plastic Matrix

A multilayer structure for bottles which has grown rapidly in use over the few years since its introduction is the combination of polypropylene, ethylene vinyl alcohol, and adhesive. It is commonly used for squeezable ketchup and mayonnaise bottles, and is also used for fruit juices and other applications where a good oxygen barrier is required. This is the plastic bottle structure chosen for investigation.

Recycled multilayer PP ketchup bottles in the form of regrind and the virgin polypropylene (Fortilene 4101, Solvay) were supplied by Continental Plastic Containers, Suisun City, California. The multi-layer bottles contain 94.50% PP, 3.75% ethylene vinyl alcohol (EVAL Solarnol DC, EVALCA), and 1.75% adhesive (Admer, Mitsui Monoply MT38). The PP is a random copolymer with ethylene.

Wood Fiber Reinforcement

In some applications, plastic materials are undesirable because they lack sufficient stiffness and are highly susceptible to creep, especially at elevated temperatures. One way to improve these properties is to combine the plastic with a filler or a reinforcing fiber. When used as a reinforcement in composite materials, wood pulp fibers possess strength and modulus properties which compare favorably with glass fibers when the density of the fibers is considered (1). Wood fibers also have distinct advantages such as lower cost, light weight, and resistance to damage during processing (2). Zadrecki and Michell (3) project cellulose fiber-thermoplastics will be introduced commercially to compete with mineral filled polymers. The advantages of thermoplastics over thermosetting resins, such as toughness enhancement of the composite and ease of processing, have spurred current research activities on thermoplastic composites (4).

A problem frequently encountered in preparing composites from wood fibers and thermoplastics is achieving adequate fiber dispersion and fiber bonding between the polar fiber and a non-polar polymer matrix. One approach which has been investigated is the use of additives to improve either dispersion or bonding (1,2,5-8).

The wood fiber chosen for the reinforcing material in this investigation was hardwood fiber (aspen), supplied by Lionite Hardboard, Phillips, Wisconsin. It was produced by a thermomechanical pulping process (TMP) and then air-dried to equilibrium at ambient conditions (23°C, 50% RH). Further information about the pulping process can be found in Simpson (9). Wood fiber produced from mechanical pulping still retains most of its lignin and natural waxes, materials which can aid fiber dispersion in nonpolar hydrocarbon polymers (2).

Wood-Fiber Plastic Composite

A combination of low cost wood fibers with recycled multi-layer plastic ketchup bottles may open up significant markets for recycling of these containers. Further, the incorporation of polar ethylene vinyl alcohol with nonpolar polypropylene may actually increase the adhesion with polar wood fibers, resulting in improved properties without any requirement for additives. In fact, recycled multi-layer PP juice and ketchup bottles have been reported to improve mechanical properties of PP homopolymers, primarily in the areas of tensile strength, elongation, flexural modulus, and impact strength (*10*). In this study, mechanical properties and dimensional stability of composites of wood fiber with virgin polypropylene (PP) were compared to those formed from regrind from multi-layer ketchup bottles (PP Reclaim).

Preparation of Composite. A Baker-Perkins Model MPC/V-30 DE, 38mm, 13:1 intermeshing self-wiping corotating twin-screw extruder was used to mix the polymer and wood fibers. Temperature of the feeder, transition, and metering zones of the extruder was 185°C. Compounder speed was 100 rpm. The polymer was added at the feeder zone while the wood fibers were added at an open port in the transition zone. The extruded material was allowed to cool to room temperature, then compression molded into plates approximately 0.125 inch thick using a Carver Model M 25 Ton laboratory press. Plates were made using three lengths of extrudate placed parallel to each other in the mold. The mold was heated at 185°C for 15 minutes under 30,000 psi of pressure, and then cooled to approximately 50°C by circulating cold water in the press for about 10 minutes. Specimens for tensile, impact, flexural modulus, creep, and water sorption were prepared according to ASTM standards (*11*).

Sample Preparation. Molded plates were cut into tensile and creep specimens (Type I dumbbell shape) using a Tensilkut Model 10-13 specimen cutter. Flexural modulus samples were cut into 6.0 inch x 0.5 inch bars using a band saw. Impact specimens were cut into 2.5 inch x o.5 inch bars and notched using a TMI Notching Cutter Model TMI 2205. Specimens for tensile, impact, and flexural modulus were made in lengthwise and crosswise direction to the extrudate. Creep specimens were cut parallel to the direction of the extrudate. Water sorption specimens were cut with a circular drill bit. All specimens were conditioned at 23°C and 50% RH for 40 hours, using Procedure A of ASTM D618-61, prior to testing.

Testing. Tensile modulus, tensile strength and elongation were measured on an Instron Tester Model 4201, following ASTM D638-87b, at ambient conditions (23°C, 50% RH). The rate of elongation was 2 inches/minute, gauge length 3.5 inches, and full scale load was 500 lbs. Sandpaper was lodged between specimens and grips to deter slippage.

Flexural modulus was tested using Method I, Procedure A of ASTM D790 on an electromechanical test frame fitted with a 20 pound load cell (United Testing System). Crosshead speed was 1.00 inch per minute, support span length was 4.0 inches, and a 16:1 span-to-depth ratio was used.

Impact testing used a TMI 43-1 Izod Impact Tester with a 5 ft-lb pendulum load, following Method A of ASTM D256-87. Averages of five to eight measurements were used to report mechanical properties.

Water sorption was determined using a 2 hour boiling water procedure (ASTM D570-81). Moisture gain was reported as an average of three measurements.

Creep extension (ASTM D2990-77) was measured by grip separation. Weights (50 lb.) were attached to the bottom of the end grips. Measurements were made at set increments up to 500 hours. Creep extension was tested in ambient and extreme (37°C, 92% RH) conditions. Extension was reported as an average of two samples and is suggestive rather than conclusive.

Results

Tensile Properties. Figure 1 shows the effect of fiber concentration on tensile strength of the composite. 30% wood fiber increased the tensile strength of the composite (compared to PP alone) in the direction of orientation, with the strength decreasing at 40 and 50% fiber. Tensile strength in the cross direction (perpendicular to orientation) was significantly lower than in the lengthwise (orientation) direction. At 30% fiber, the PP Reclaim composite was superior to the PP composite, while at higher fiber concentrations properties were much the same. For the unreinforced matrix material, the virgin PP exhibited a somewhat higher tensile strength than the PP Reclaim.

Elongation at break decreased with increasing fiber concentration, with the PP Reclaim showing much greater elongation at 30% fiber than the PP composite (Figure 2). These differences also decreased as fiber concentration increased. Elongation was greater in the direction of orientation than in the cross direction. For the unreinforced material, PP showed 690% elongation at break, compared to 215% elongation at break for PP Reclaim.

Young's modulus as a function of fiber concentration appears in Figure 3. Tensile modulus increased with an increase in fiber concentration. Tensile modulus was also highest in the direction of orientation, as was tensile strength. Tensile modulus for the PP Reclaim composite was lower than for the PP composite, at all fiber concentrations. The unreinforced materials did not differ significantly in tensile modulus.

Flexural Modulus. The flexural modulus for the PP Reclaim composite increased with increasing fiber content (Figure 4). The PP composite showed somewhat varying results, but still with a trend towards increase with increasing fiber content. At 40 and 50% wood fiber, in the direction of orientation, the PP Reclaim composite had a higher flexural modulus than the PP composite. Values in the cross direction were nearly the same for the PP

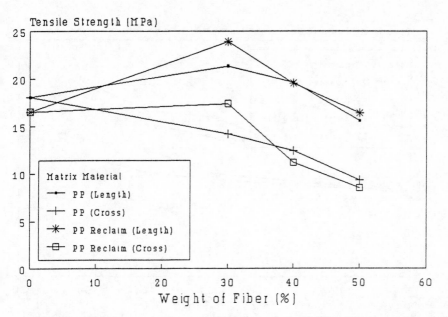

Figure 1. Tensile strength as a function of wood fiber content and orientation.

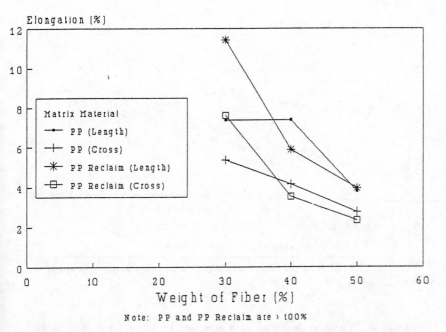

Note: PP and PP Reclaim are > 100%

Figure 2. Elongation as a function of wood fiber content and orientation.

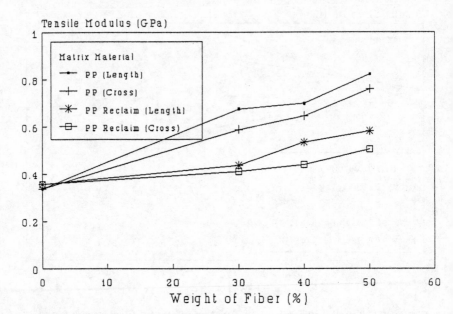

Figure 3.   Tensile modulus as a function of wood fiber content and orientation.

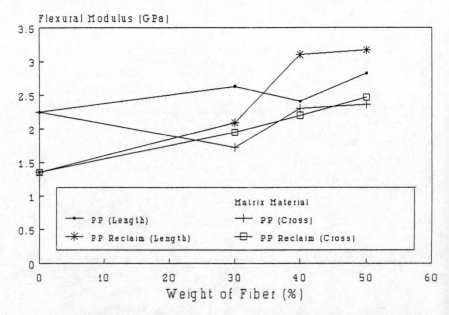

Figure 4.   Flexural modulus as a function of wood fiber content and orientation.

and the PP Reclaim. For the unreinforced materials, the PP had a significantly higher flexural modulus than the PP Reclaim.

Izod Impact. As fiber concentration increased, the notched Izod impact strength also generally increased (Figure 5). This behavior is in contrast with other studies which indicate a decrease in Izod fracture energies in proportion to the added fiber (2, 12). For the unreinforced material, the impact strength of the PP Reclaim was somewhat higher than that of the virgin PP. For the composites, the PP Reclaim appeared superior at 40 and 50% fiber, but not at 30% fiber.

Dimensional Stability. Extreme environmental conditions severely affected creep extension in both composite structures in comparison to ambient conditions (Table I). The PP composite exhibited poor dimensional stability at 50% fiber content and broke after 20 hours in extreme conditions. At 40% wood fiber content in extreme conditions, the elongation of the PP composite was 140% greater than that of the PP Reclaim composite. These results suggest that the structural materials in the PP Reclaim provide longer retention of strength under wet conditions. For the unreinforced material, the PP Reclaim also exhibited lower creep than the virgin PP.

Water sorption increased with increasing fiber concentration, as shown in Table II. Differences between the PP and the PP Reclaim were not clear.

Table I. Effect of Fiber Content on Creep Extension (500 h)

| Matrix and Condition | Increase in Length (mm) | | | |
	No Fiber	30%	40%	50%
PP - Ambient	1.41	0.60	0.67	0.66
PP Reclaim - Ambient	1.03	0.44	0.34	0.44
PP - Extreme	3.20	2.39	4.60	
PP Reclaim - Extreme	3.06	2.69	1.92	1.93

Note: 50% PP-wood fiber composite failed after 20 h.

Table II. Effect of Fiber Content on Water Absorption

| Matrix | Increase in Weight (%) | | | |
	No Fiber	30%	40%	50%
PP	0.10	1.65	2.79	3.76
PP Reclaim	0.24	1.41	2.36	3.88

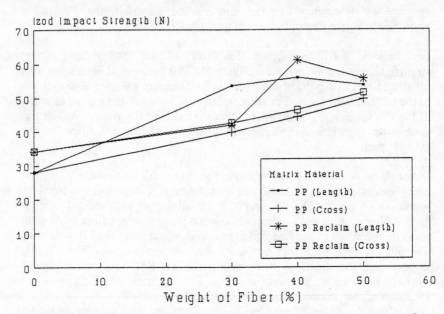

Figure 5.   Impact strength as a function of wood fiber content and orientation.

Conclusions

The PP Reclaim-wood fiber composite exhibited improved mechanical properties compared to the PP composite. Increase in wood fiber content and orientation of the fibers improved the mechanical properties for both composites, except for tensile strength and elongation at break, both of which decreased at fiber contents above 30%. The PP Reclaim composite also exhibited significantly less creep, especially under severe environmental conditions.

An increase in interfacial adhesion due to the polar groups contained in the ethylene vinyl alcohol and in the adhesive is a likely explanation for the improved properties. The results demonstrate that this multilayer bottle structure can provide a matrix for wood fiber composites which is actually more desirable than that of the major resin alone. Therefore recycling of these bottles into wood fiber composites may be both a viable and a valuable option.

Acknowledgments

The authors would like to thank the State of Michigan Research Excellence Fund, the Composite Materials and Structures Center at Michigan State University, and the USDA for their financial support. We would also like to thank Continental Plastic Containers and Lionite Hardboard for the materials used for this study.

Literature Cited

1. Woodhams, R. T.; Thomas, G.; Rodgers, D. K. *Polym. Eng. Sci.* **1984**, *24*, 1166-1171.
2. Raj, R. J. B.; Kokta, B. V.; Maldas, D.; Daneault, C. *Polym. Comp.* **1988**, *9*, 404-411.
3. Zadorecki, P.; Michell, A. J. *Polym. Comp.* **1989**, *10*, 404-411.
4. Bigg, D. M.; Hiscock, D. F.; Preston, J. R.; Bradbury, E. J. *Polym. Comp.* **1988**, *9*, 222-228.
5. Bataille, P.; Allard, P.; Cousin, P.; Sapieha, S. *Polym. Comp.* **1990**, *11*, 301-304.
6. Bataille, P.; Ricard, L.; Sapieha, S. *Polym. Comp.* **1989**, *10*, 103-108.
7. Kokta, B. V.; Maldas, D.; Daneault, C.; Beland, P. *Polym. Comp.* **1990**, *11*, 84-89.
8. Selke, S.; Yam, K.; Nieman, K. *ANTEC'89*, Technical Papers, Society of Plastics Engineers, Inc.: Brookfield, CT, 1989, Vol. 35, pp. 1813-1815.
9. Simpson, R. J., *Composite Materials from Recycled Multi-layer Polypropylene Bottles and Wood Fibers*, MS Thesis, Michigan State University, 1991.
10. *Plastics World* **1990**, *48* (Aug.), 61.
11. *Annual Book of ASTM Standards, Section 8: Plastics*; American Society for Testing and Materials: Philadelphia, PA, 1988.
12. Raj, R. G.; Kokta, B. V.; Daneault, C. *Sci. Eng. Comp. Mtls.*, **1989**, *1*, 80-98.

RECEIVED March 9, 1992

Chapter 19

Blends of Nylon 6 and Polypropylene with Potential Applications in Recycling

Effects of Reactive Extrusion Variables on Blend Characteristics

S. S. Dagli, M. Xanthos, and J. A. Biesenberger

Polymer Processing Institute, at Stevens Institute of Technology, Castle Point, Hoboken, NJ 07030

As part of efforts to develop recycling technologies through melt reprocessing of commingled plastic streams (e.g. fishing gear or carpeting) in compounding extruders the effects of process parameters on the characteristics of a model blend containing nylon 6 (85% by wt.) and polypropylene are reported. Twin-screw extruder runs and preliminary batch mixer experiments showed that the presence of an acrylic acid modified polypropylene acting as reactive compatibilizer affects significantly flow properties and promotes finer morphology and improved blend properties. Other reactive extrusion process variables such as screw speed, residence time, venting and feeding sequence also affect blend properties, but to a lesser extent. Results are interpreted in terms of parameters such as components viscosity and thermal stability, reaction kinetics and mixing efficiency.

Among the various marine plastic debris, discarded plastic fishing gear is a major pollutant, adversely affecting marine life. Virtually any marine species could die from entanglement in fish nets lost or abandoned. Recent estimates of the U.S. Academy of Sciences bring the amount of fishing gear lost in the sea (and often washed up on the beaches) to about 135,000 metric tons per annum (*1*). During a year long effort in collecting discarded fish nets and identifying the polymers used in making them (*2*), it was found that the majority of nets were made of nylons and polyolefins, whereas ropes and lines were mostly polypropylene. At present, plastic fishing gear is not recycled in the U.S. To help in the efforts of developing a recycling technology for the plastic fishing gear, a basic study on blends of nylon 6 (N6) and polypropylene (PP) was undertaken. Mixtures of N6 and PP are also encountered in recycling of automotive carpeting and, thus, the findings of this work could also be applicable in this area.

N6 and PP are immiscible polymers. When blended together they form an incompatible blend. Incompatible blends do not have useful properties. To improve the properties of such a blend, it would be necessary to compatibilize it with a compatibilizer added as a third component, or it can be formed in-situ during blend compounding by using suitable functionalized components. Compatibilizing

0097–6156/92/0513–0241$06.00/0

improves the short and long-term mechanical properties of the blends. This is achieved mainly due to an improvement in the morphology (finer and more stable). The in-situ formation of compatibilizers for polyblends is rapidly gaining popularity. In the last few years, more and more cases have been reported where advantage is taken of the presence of reactive groups to form a graft or block or random copolymer, which could act as a compatibilizer for a mixture of two or more polymers (3-5).

In the early 1970's, Japanese workers (6-9) reported the use of maleic anhydride grafted polypropylene (PP-g-MAH) to improve the dispersibility and mechanical strength of N6 / PP blends. Since then, several studies were reported on laminates of N6 and PP and, in particular, on improving their adhesion by the use of modified PP or other modified polar thermoplastics in the packaging and automotive industries (10-12). Again in the mid and late 1980's, additional work was reported where modified PP, ionomers, or other polar substances were used to improve the compatibility of N6 / PP blends (13-16). There were also a few studies published on the composition dependence of the properties of N6 / PP blends without any third component (17,18).

In the present study, an acrylic acid grafted modified polypropylene (PP-g-AA) is used with a blend of nylon-6 (N6) and polypropylene (PP). In addition to strong specific interactions such as H-bonding, the amine end groups of N6 are expected to react with the acid groups of PP-g-AA with the evolution of water to form a graft copolymer in-situ. Under processing conditions, PP-g-AA may also form an anhydride (19) so an anhydride/amine reaction is also possible. Both these reactions in a simplified form are shown in Figures 1 and 2. Both reactions can form a graft copolymer by grafting PP molecules onto N6 molecules. Thus, a copolymer having parts common with both the phases could be formed which could act as a compatibilizer. PP-g-AA could have more than one acid group on each graft since polyacrylic acid oligomers are usually grafted on the PP backbone. This would mean multiple reactive sites and, hence, multiple graft reactions as shown in Figure 1. This could lead to some cross-linking of N6 and to anhydride formation as mentioned earlier. Besides these reactions, an acid (or anhydride) /amide reaction (acid or anhydride groups reacting with the amide groups on the nylon backbone) forming an imide is also possible, though not very likely (20,21) as the amine end is much more reactive, when compared with the amide groups. Also such a reaction could cause the N6 molecule to split, as is the case for the reaction with the anhydride groups. Lawson et al. (20) have carried out an extensive analysis of a similar reactive system and have not found any signs of reduction in molecular weight of the N6 thus, ruling out such a reaction completely. There could be some ungrafted polyacrylic acid present in PP-g-AA (22). This could also participate in reactions with N6 and form H-bonds.

Parameters Controlling Process Requirements.   There are some important process parameters which would have significant effect on reactive polymer blending experiments. The following parameters were considered in the present study :

Viscosities of the components: Mixing two polymers with vastly different melt viscosities is a difficult challenge.   Various studies (23,24) on mixing components with different viscosities show that the mixing is the most efficient (easiest to attain the lowest dispersed size) when the viscosity ratio of the components is close to one.   In this study the aim was not to investigate the effects of viscosity ratio during reactive extrusion, and, thus, a proper selection

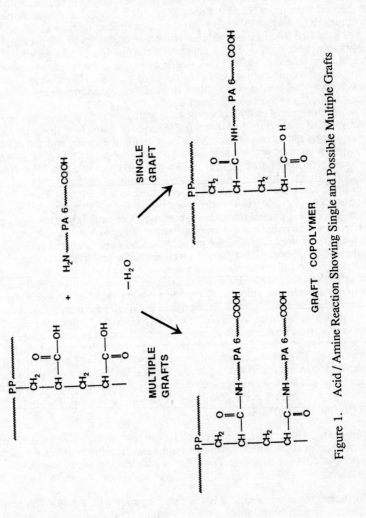

Figure 1. Acid / Amine Reaction Showing Single and Possible Multiple Grafts

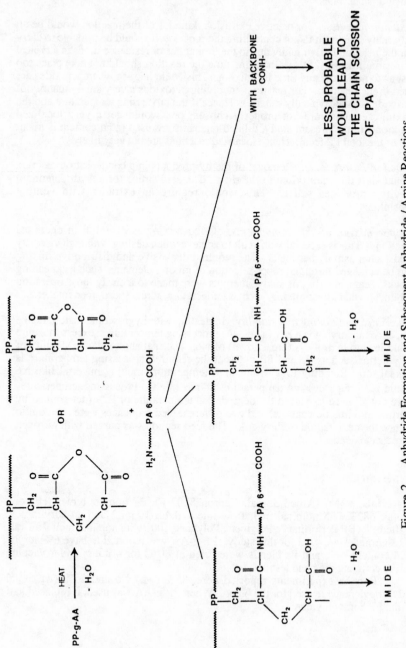

Figure 2. Anhydride Formation and Subsequent Anhydride / Amine Reactions

of grades of components was necessary. The criterion for selecting the grade of PP used in this study was that its viscosity be similar to that of N6 over a wide range of shear rates.

Reaction Kinetics: For reactive extrusion, kinetics of the reaction would be an extremely important factor controlling the process. It should be possible to carry out the reaction fairly rapidly due to the limitations of residence times in a typical extrusion process (25). At the same time the reaction should not take place too fast as to not allow any time for mixing. This could happen when peroxides are used in the reaction. For reactions involving peroxides a very small quantity of peroxide is used (typically < 0.5%). Hence if the processing temperature and the mixing geometry are not properly chosen, one would get a very localized reaction. In the present study, diluted reactants are used (acid content is about 2% of the total PP feed). Hence localized reactions are not very likely.

Reaction By-products: Removal of the by-products in a condensation reaction would shift the equilibrium to the right. The acid/amine (or anhydride/amine) reaction produces water. This would require an extruder with venting capabilities.

Proper mixing of the components: Good mixing is essential in chemical reactions. This is especially difficult to achieve when dealing with high viscosity feeds, such as polymer melts. The required dispersive and distributive mixing can be achieved through proper selection of mixing elements, such as kneading blocks, gears, etc. In this study, attempts were made to identify those operating parameters that affect mixing, such as temperature, screw speed, feed rate, etc..

Materials conditioning and stability: In dealing with hygroscopic materials like N6, adequate drying is a key factor. At processing temperatures, excess moisture could lead to hydrolysis (breakdown) of N6. Thermal and oxidative stability of polyolefins such as PP and PP-g-AA at the required processing temperature is crucial too. Blending PP's with a high temperature melting material like N6 would require processing temperatures above 230^0C. Processing temperatures were not allowed to exceed the degradation temperature of PP's (determined by thermal gravimetric analysis) and two different feed sequences were tried out to reduce the degradation of PP-g-AA. The feed section was purged with nitrogen to reduce oxidation too.

Experimental

Materials. N6 (Allied-Signal, Capron 8207-F), PP homopolymer (Himont, Profax 6823 with nominal melt flow rate = 0.4 g/10min.) and an acrylic acid grafted PP (BP Performance Polymers, Polybond 1001 with nominal melt flow rate = 12 g/min.) were used for this study. PP-g-AA was reported to have 6% acrylic acid content. N6 and its blends were dried at 80°C for 4 hours under vacuum before compounding and testing.

For the experiments reported here, 85% (wt.) N6 and 15% (wt.) total polypropylene were used for the blend. When PP-g-AA was used, it replaced half of the PP homopolymer.

Processing.

A. Batch Compounding: The batch compounding experiments were carried out in intensive batch mixers (Haake-Buchler Rheocord System 40 and C.W. Brabender PL2000 Computerized Plasti-corder) equipped with roller blades. The operating temperatures were in the range 230-250°C and rotor speeds in the range 20-60 RPM. A nitrogen blanket was used during most of the compounding experiments to prevent oxidative degradation. The polymers were dry blended prior to their addition to the mixer. A Factorial design approach (26) was used to evaluate the importance of various variables varied at two levels as listed below :

Variable	Level 1	Level 2
Temperature, °C	230	250
Rotor speed, RPM	20	60
PP-g-AA	Absent	Present

B. Extrusion Compounding: Continuous compounding was carried out in a Werner & Pfleiderer 30mm corotating intermeshing twin screw extruder (ZSK 30). During compounding experiments, two different feed sequences were used :
- a) Combined feed : All the components fed at the same location (hopper of the extruder)
- b) Split feed : With PP fed in the hopper and N6 (or a dry blend of N6 and PP-g-AA) fed as a solid sidestream (at about 20% of the total length from the hopper).

When just PP homopolymer was fed in the hopper, temperature settings of the first zone were 165°C, increasing to 220°- 240°C for the zones thereafter. In the case of combined feed temperature, settings were in the range 220° - 240°C throughout. The melt temperature (measured just before the die) ranged from 235° to 255°C. In the case of split feed, nitrogen gas was fed in the second port to reduce the oxidative degradation of the exposed molten PP. The screw profile is shown in Figure 3. One short kneading section was placed before the second feed. A large section containing kneading blocks separated by a conveying element before the vent section provided the necessary intensive mixing and melt seal for the vacuum. The same screw configuration was used for both the feed sequences. This was done to avoid an additional variable viz. the screw profile entering the study. The material was extruded using a single strand rod die of 4.8 X 10^{-3}m diameter. The extruded strand was water-cooled and pelletized using a Killion pelletizer. Under the above process conditions, typical minimum residence times were estimated to be 2 to 3 minutes.

A saturated fractional factorial design (26) was used to evaluate the importance of various variables varied at two levels as listed below :

Variable	level 1	level 2
Feed rate, kg/hr (lb/hr)	4.5 (10)	9.0 (20)
Temperature settings, °C	235	255
Screw Speed, RPM	200	400
Feed sequence	Combined	Split
PP-g-AA	Absent	Present
Vacuum	Not Applied	Applied

C. Injection Molding: Materials compounded in the twin screw extruder were injection molded after drying. A Van Dorn (40 ton) injection molding machine with a 4 cavity-ASTM test specimen mold was used.

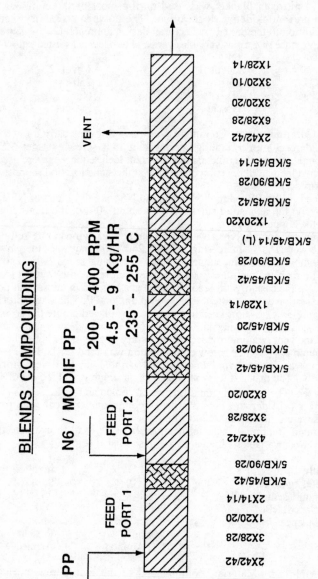

Figure 3. 30 mm Co-rotating Intermeshing Twin Screw Extruder Setup

Typical molding conditions were as follows :

Barrel Temperature °C	250 - 260
Mold Temperature °C	95 - 100
Screw Speed, rpm	200
Injection Pressure, psi	720

Characterization:

A. Rheological characterization: (a) Melt Flow Rate: Melt flow rate (MFR) of the components and the compounded blends were determined using a Tinius Olsen Plastometer with digiset controller (Model: U E-4-78). For these tests, N6 testing conditions i.e. 235°C and 2.16 kg load, were used. All N6 and N6 containing samples were pre-dried for four hours under vacuum at 80°C.

(b) Shear Viscosity: - Instron capillary rheometers (Floor models: TFD and 4204) were used for determining the shear viscosity of the components at various temperatures in the processing range with a capillary die of $D= 7.6 \times 10^{-4}$ m and $L/D=33$. Rabinowitsch's equation was used to correct the shear rate values.

B. Infrared Spectroscopy: For infrared (IR) spectroscopy, very thin films were prepared by compression molding components and blends. A Perkin-Elmer IR spectrophotometer model 983 was used. All samples were scanned in the 4,000 to 400 cm^{-1} range.

C. Solubility Test: The presence of graft reaction product between N6 and PP-g-AA was tested by applying Molau's test (27). An 80% formic acid solution was used for this test. Formic acid is a solvent for N6 and non-solvent for polypropylenes. In the absence of grafting, submerging the product in formic acid should result in a clear solution, with the PP phase floating on top. With grafting, the PP should not be completely separable, thereby resulting in turbidity.

D. Mechanical Testing: Tensile properties of injection molded test specimens were determined in a Tinius Olsen Tensile Testing Machine (LO-CAP Universal Testing Machine) at 0.05 m/min. The samples were not conditioned before testing (tested "as is"). Reported values are averages of 5 to 6 measurements.

E. Morphology: (a) Sample Preparation: Samples were cut at room temperature on a Reichert-Jung Ultracut E Microtome using a glass knife. Thin slices were collected for optical microscopy. Samples with smooth surface for scanning electron microscopy (SEM) were also prepared by microtoming. For the SEM study PP was dissolved by immersing samples in boiling xylene.

(b) Microscopy: Thin sections were observed in transmitted light on a Zeiss Universal Optical Microscope. For scanning electron microscopy (Joel Model 840), after chemically extracting the minor phase, samples were coated with an alloy of gold and palladium using a sputter (Model 13131, SPI) with a magnetron cathode. Photomicrographs from SEM were analyzed for average domain size, percentage area, roundness etc. using an image analyzer (Quantimet Q-10).

Results and Discussion

Shear Viscosity curves for the blend components (N6 and PP) at 240 °C are shown in Figure 4. The same Figure also shows the viscosity ratio (VR) as a function of the shear rate. The viscosities of the components are not vastly different in the extrusion processing range of shear rates and this should facilitate mixing. Furthermore the viscosity of N6 is slightly higher than that of PP, which would probably assist in dispersing small amounts of PP in the N6 matrix.

Experiments in the Intensive Batch Mixer: (a) Effects of Process Parameters: To estimate the importance of the parameters studied (temperature, rotor speed, and presence of PP-g-AA) equilibrium mixing torque (t) and slope of the input energy curve (de/dt) were examined. de/dt which is the power input, should be a good indicator of any structural changes taking place. Such changes would include chain extension / graft reactions between acid and amine groups, leading to an increase in viscosity, which should be reflected in a faster increase in the energy input and higher t values. The "effects" were calculated for each parameter. The results indicate:

1) For a given blend increasing temperature had a negative effect. The values of t and **de/dt** are shown in the Figure 5 for the binary (N6/PP) and the ternary (N6/PP/PP-g-AA) blends at two temperature levels. Increasing temperature lowers both t and **de/dt**. This could be explained by the drop in viscosities due to increase in temperature and possible thermal degradation of the polypropylene resins.
2) The presence of the modified PP has a very significant effect on both de/dt and t. (See Figure 5). This suggests the occurrence of structural changes.
3) For ternary blends increasing the rotor speed has a very strong effect on **de/dt** (data not shown).
4) Temperature and presence of PP-g-AA did not show any correlation.

To summarize, for the N6 and PP blend, the presence of PP-g-AA emerged as a variable having the strongest effect on both de/dt and t. The presence of PP-g-AA gave much higher **de/dt** and t values. In the case of ternary blends, increasing the rotor speed seemed to promote reactions between blend components as indicated by the higher de/dt and t values. Results of the batch experiments also indicate that the compatibilizing reaction is fast enough to be carried out in a continuous fashion in an extruder.

Solubility Test: The addition of the ternary blends in formic acid resulted in a turbid solution, thereby suggesting the formation of a graft copolymer. The addition of binary blends resulted in a clear solution with PP floating on the top.

Infrared Spectroscopy: PP-g-AA showed a characteristic absorption peak at wavelength 1700 - 1725 cm-1 (free acid carbonyl - COOH). The carbonyl peak from N6 backbone (-CONH-) is very strong and it partially overlaps the free acid carbonyl peak. In the unreacted blend the free acid peak appears only as a shoulder. This shoulder is not visible in the reacted blends.

Extrusion Experiments: The variables considered for the extrusion experiments were feed rate, temperature settings, screw speed, addition of PP-g-AA and vacuum; each was varied at two levels as described in the experimental

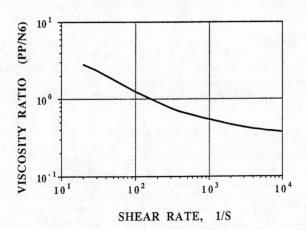

Figure 4.   Viscosity Ratio (PP/N6) at 235°C

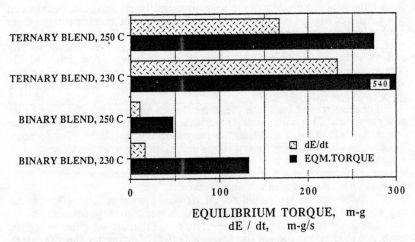

Figure 5. Intensive Batch Mixer Data—Equilibrium Torque and Slope of the Energy Input Curve for the Binary and Ternary Blends at Two Temperatures.

part. Product properties like morphology (domain size), melt flow rate and tensile strength and elongation were used to judge the effects of these parameters on the extrusion process. Some additional experiments were carried out to judge the isolated effect of a parameter. The effect of these variables on the above mentioned product properties are expressed qualitatively in Table 1 as "high", "mod." (moderate) and "low". A negative sign indicates a negative effect i.e. increasing a variable (or moving to the second level of the variable) results in a decrease in the measured property.

For example: (a) Addition of PP-g-AA (second level for that variable) decreases MFR of the blend significantly. This is expressed as "-high" in Table 1. (b) An increase in screw speed results in a moderate drop in the domain size. This is expressed as "-mod." (c) An increase in the feed rate results in a marginal increase in the tensile strength at yield. This is expressed as "low" .

The overall effect of adding the modified PP was very significant. It reduced the MFR significantly as shown in Figure 6 (in agreement with the previous **t** and **de/dt** increase results) and resulted in a noticeable drop in the domain size. The domain size used in calculating the effects was the mean of average longest and shortest sizes measured at two locations of an extruded strand as illustrated in Figure 7. SEM photomicrographs of a representative blends, with and without PP-g-AA are shown in Figure 8. The presence of modified PP also has a significant effect on the tensile properties, improving the tensile strength (both @ yield and @ break) as much as 35% in some cases. Figure 9 shows tensile strength data.

The feed sequence had a moderate to no effect on the blend properties except in the case of tensile strength @ yield which was significantly affected. Consistently, when components were fed separately, the resulting blends exhibited lower tensile strength. For the binary blend, feeding the components together should reduce the chances of oxidative degradation of PP. This could explain the higher tensile properties when the components are fed together. For the ternary blend in addition to the reduced chances of oxidative degradation of the polypropylenes, longer reaction time and higher stresses generated during solids melting would contribute to the higher tensile properties observed when the components were fed together.

For the ternary blends, the presence of vacuum seemed to result in better tensile properties. This observation supports the hypothesis that the removal of the by-product would promote acid/amine or anhydride/amine reaction. The presence of vacuum produced also finer morphology in the case when the components were fed separately (See Figure 10). Such an effect on morphology was not observed when the components were fed together. In general, the presence of vacuum had little effect on the properties of the binary blends.

Temperature on its own did not seem to have any significant effect on the blend properties probably because of the narrow temperature range covered.

The screw speed had a moderate effect on the domain size. Increasing the screw speed reduced the domain size (expected effect of increasing shear stresses). It also reduced the MFR moderately. The tensile strength and % elongation @ yield were affected too. Higher screw speed runs yielded lower tensile strength @ yield (due to possible degradation of the components) and higher % elongation. Finally, changing the feed rate did not have appreciable effect on the measured properties.

TABLE 1. RESPONSE OF PROPERTIES TO CHANGES OF PROCESSSING PARAMETERS IN TWIN SCREW COMPOUNDING

	Feed Rate	Temp- erature	Screw Speed	Feed Sequence	Presence of PP-g-AA	Vacuum
Melt Flow Rate	-low	low	-mod.	-low	-high	mod.
Domain Size	low	low	-mod.	low	-high	low
Tens. Strength. @ Yield	low	-low	-mod.	-high	high	low
Tens. Strength @ Break	low	-low	-low	-mod.	high	low
% Elongation @ Yield	-low	-low	mod.	mod.	-high	low
% Elongation @ Break	-low	-mod.	-low	mod.	-high	mod.

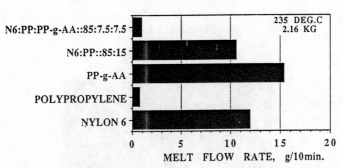

Figure 6.   Melt Flow Rate Data for the Blends and Components

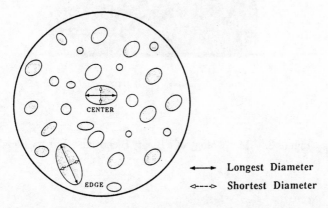

Figure 7.   Domain Size Measurement Showing Measurement Locations

TERNARY BLEND

N6 / PP / PP-g-AA

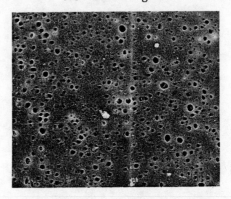

BINARY BLEND

N6 / PP

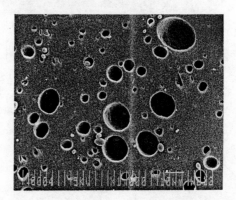

— 10 μm

Figure 8. Morphology of a Binary Blend and a Ternary Blend

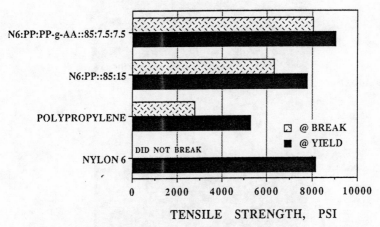

Figure 9.   Tensile Strength Data for the Blends and Components

TERNARY BLEND (N6/PP/PP-g-AA)

VACUUM NOT APPLIED

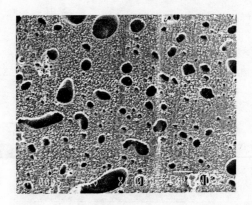

VACUUM APPLIED

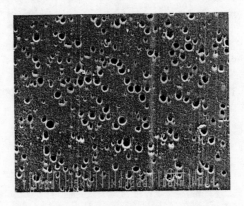

—— 10 μm

Figure 10. Effect of Venting on the Morphology of a Ternary Blend

Conclusions

The presence of PP-g-AA results in noticeable differences in the flow properties, morphology and tensile properties of a N6 / PP blend. This indicates that structural changes are taking place such as the formation of graft copolymer. More direct evidence for the presence of such copolymer was provided by the results of the solubility test. From a processing viewpoint, trends observed in a laboratory scale batch processing were also observed during extrusion compounding. Screw speed, residence time, and venting had noticeable effects on the reactive extrusion process. Combined feed resulted in better product properties. Temperature and feed rate were not found to be significant parameters in the range tested. The results of this work could be applicable to the development of reprocessing technologies for potentially recyclable blends of Nylon / PP from fishing gear, automotive carpeting and other sources.

Literature Cited

1.   Lautenberg, F. (Senator), *Introduction of "Plastic Pollution Control Act of 1987", U.S. Congressional Record-Senate, S269S*, March 3, **1987**.
2.   Dagli, S. S.; Kelleher, P. G.; Monroy, M.; Patel, A. ; Xanthos, M., *Adv. Polym. Tech.* **1990**, *10*, pp.125-34.
3.   Xanthos, M.; Dagli, S. S. *Polym. Eng. Sci.*, **1991**, *31*, pp.1-7.
4.   Gaylord, N. C. *J. Macromol. Sci. Chem.*, **1989**, *A26*, pp.1211.
5.   Frund Jr., Z. N. *Plastics Compounding*, **1986**, *Sept./Oct.*, *9*, pp.24.
6.   Ide, F.; Kodama, T.; Hasegawa, A. *Kobunshi Kagaku,* **1972**, *29, 4*, pp.259-64 (Engl. Abst.).
7.   Ide, F.; Rodama, T.; Hasegawa, A. *Kobunshi Kagaku*, **1972**, *29, 4*, pp.265-69, (Engl. Abst.).
8   Komatsu, F.; Kaeriyama, A. *Muroran Kogyo Daigaku Kenkyu; Horoku*, **1972**, *7*, 3, pp.719-38 (Engl. Abst.).
9   Ide, F.; Hasegawa, A. *J. Appl. Polym. Si.*, **1974**, *18*, pp.963-76
10   Mitsui Petrochemical Industries, Ltd.; JP 60,110,740, **1985**. CA <u>103</u>: 161550g, **1985**.
11   Katsura, S. Mitsui Petrochemical Industries, Ltd.,; JP 61 28,539, **1986**. CA <u>105</u>: 44048n, **1986**. ‛
12   Sakuma, M.; et al, Tonen Sekiyu Kagku, Inc.; JP 62,158,739, **1987**; CA <u>180</u>: 95483v,**1988**.
13   Van Gheluwe, P.; Favis, B. D. *Polym. Mater. Sci. Eng.*, **1988**, *58*, pp.966-970.
14   Willis, J. M.; Favis, B. D. *Polym. Eng.Sci.,* **1988**, *28*, pp.21.
15. Van Gheluwe, P.; Favis, B. D.; Chalifoux, J. P. *J Matl. Sci.,* **1988**, *23*, pp.3910-3920.
16. Scholz, P.; Froelich, D.; Muller, R. *J. Rheol.,* **1989**, *33*, pp.481-99.
17. Hayashida, K.; Yoshida, T. *Kyoto Kogei Seni Daigaku, Sani Daigaku Senigakubu; Gakujutsu Hororu*, **1979**, *9*, 1, pp.65-72 (Engl. Abstr.).
18. Laing, B.; White, J. L.; Spruiell, J. E.; Goswanni, B. *J. Appl. Polym. Sci.,* **1983**, *28*, 6, pp.2011-32.
19. Jin, Y.; Huang, R. Y. M. *J. Appl. Polym. Sci.,* **1988**, *36*, pp.1799-1808.
20. Lawson, F. D.; Hergenrother, W. L.; Matlock, M. G. *Polymer Preprints, ACS,* Sept., **1988**, *29*, pp.193-94,
21. Hergenrother, W. L.; Matlock, M. G.; Ambrose, R. J. U.S. Pat. # 4,508,874, **1985**.

22. Adur, A. Personal communication, BP Performance Polymers, Inc., Hackettstown, NJ, Feb., **1991**.

23. Starita, J. M. *Trans. Soc. Rheol.,* **1972**, *16*, pp.339-67.

24. Favis, B. D.; Chalifoux, J. P., *Polym Eng Sci.,* **1987**, *27*, pp.1591-1600.

25. Todd, D. B. *Polym. Prep., A.C.S.,* **1988**, *29*, 1, pp.563.

26. Box, G. P. E.; Hunter, W. G.; Hunter, J. H. *Statistics for Experimenters,* John Wiley & Sons, New York, NY, **1978**.

27. Molau, G. E. *J. Polym. Sci., Part A,* **1965**, *3*, pp.1267-78.

RECEIVED May 1, 1992

Chapter 20

Properties of High-Density Polyethylene from Postconsumer Recycled Containers

Philip S. Blatz

Du Pont Polymers, Experimental Station 323/210,
Wilmington, DE 19880−0323

High density polyethylene obtained from recycled post consumer containers can be readily fabricated into useful items by conventional injection molding and extrusion blow molding processes. The properties of moldings produced from the recycle material have properties very similar to virgin material. Commercial items are being produced and sold using the recycled HDPE. As the volume of recycled HDPE increases, the types of products made from these materials will increase and will be favorably received by the consumer.

A major issue facing municipalities today is the environmentally safe and cost-effective disposal of solid waste. Most of the waste, about 80%, goes into landfill, 10% is burned through incineration and only 10% is recycled. Although, the perception by the public is that plastics are a major contributor to the solid waste stream, the facts are different. In 1988, plastics made up only 7.3% by weight or about 18% by volume of the material going into a landfill, with paper and paperboard making up 35.6% and yard wastes contributing 20.1%.

And again although the public perception is that plastics packaging is the major contributor to the solid waste problem, the facts are different. Paper and paperboard packaging makes up almost 50% and glass 25% by volume of the waste stream from packaging, whereas plastics contributes only about 13% by volume. Still, because the public perception is that plastics are a major contributor to the waste problem, the plastics industry is taking steps to change that perception. The public must be educated to understand that plastics are not the major contributor to the solid waste problem and that plastics can and are being recycled.

0097−6156/92/0513−0258$06.00/0

Plastic Recycling Alliance

One significant step that The Du Pont Company has taken is the formation of the Plastic Recycling Alliance, for the purpose of recycling plastic bottles. To date, recycling plants have been established in Philadelphia and Chicago. Each plant can recycle 20 MM lbs of plastic each year.

The plants initially are accepting post consumer containers based on polyethylene terephthalate (PET) and high density polyethylene (HDPE). The three major containers that the recycling plants process are soda bottles made from PET, milk jugs from natural HDPE, and colored detergent bottles. There are five products from the plants based on these three types of containers: clear uncolored recycle PET flake; green recycle PET flake; natural unpigmented HDPE from the recycled milk jugs; mixed color recycled HDPE detergent bottle flake; and HDPE mixed color recycle soda bottle HDPE base cup flake. The detergent bottle flake and the base cup flake also contain the polypropylene caps from the bottles. There is demand for the clear recycle PET, the natural recycle HDPE, and of course companies selling detergents like the mixed color recycle detergent HDPE. There is not much demand at present for the green PET flake or the predominantly black HDPE base cup flake.

In order to understand that recycling the PET soda bottle is not a trivial process, it must be pointed out that the soda bottle is made up of at least 6 different materials. Table I shows the materials that go into a typical 2 liter soda bottle container.

Table I
Components of a Typical 2-Liter Soda Bottle

Description of Component	Polymer	Weight in Grams	Percent of Total
Stretch-Blown Bottle	PET	49.8	68.7
Base Cup	HDPE	17.6	24.4
Cap	PP	2.6	3.6
Label	PP + Ink	1.7	2.4
Glue for Base	Hot Melt based on E/VAc Copolymer	0.4	0.6
Gasket in Cap	Butadien/Styrene Block Copolymer	0.2	0.2
Glue for Label	Unknown	0.1	0.1

The PET stretch blown bottle and the HDPE base cup make up about 93% of the total with the two materials present in a ratio of about 74 to 26. It is seen that the container also has a polypropylene cap and label, a butadiene/styrene block copolymer in the gasket, and an ethylene vinyl acetate copolymer based hot melt for attaching the PET container to the HDPE base cup. We have not analyzed the adhesive used to attach the label. It may also be based on an ethylene/vinyl acetate copolymer. But if a stretch blow molded bottle that is clear is to be produced from the clear recycled PET material it is not easy to ensure that all of the glue and gasket materials are removed from the PET flake. It doesn't take much of one of the contaminants to produce a hazy or cloudy bottle.

Properties of HDPE Recycle Products

This paper discusses the properties of the three HDPE products that come out of the recycle facilities. All three products have been injection molded and the recycle milk jug and detergent bottle flake have been blow molded. The properties of the injection moldings will be discussed first. The three HDPE products have been extrusion blended and pelletized and then injection molded to produce 1/8" flex bars and 1/8" tensile bars from recycled polymer. Samples obtained from the facilities over a period of several weeks were processed. Mechanical properties of the moldings are shown in Table II.

Table II
Properties of Injection Molded
Recycle High-Density Polyethylene

Property	Recycle Milk Jug	Recycle Mixed Detergent Bottle	Recycle Soda Bottle Base Cup
Color	Off White	Olive Drab	Black
Melt Index G/10 Min	0.40-0.60	0.35-0.45	5.0-20.0
Tensile Strength PSI	3000-4500	2500-4000	1500-2500
Elongation %	500-700	260-440	5-10
Flex Modulus KPSI	100-110	60-95	120-135
Notched Izod Ft-Lb/In	19-21	11-15	0.3-0.6

The variation of properties arises from the different resins that were used to produce the containers as well as the degree of contamination present in the products exiting the facility. The data in Table II can be summarized as follows:

1. The three HDPE products have decidedly different colors. Only the resin from the milk jugs have a natural but slightly off white color. The detergent bottle resin is a mixture of many bottles from a rainbow of fluorescent colors. Melt blending the flake at conventional melt temperatures produces an olive drab to tan color with no fluorescence. The base cup resin is a mixture of mainly black and green with minor amounts of other colors. The color will vary from dark green to black depending on the relative amounts of the two major colors.

2. As both of the resins used for the milk jugs and detergent bottles are blow molding grade, their melt indices are less than 1. The variation in the melt index of these resins will depend on the specific melt index of the virgin resins as well as the level of residual milk or detergent products which are absorbed in the wall of the bottles. The large variation in the melt index of the base cup resin is caused by the presence of the other components of the soda bottle container such as the polypropylene cap and label as well as the adhesives. As the PET flakes are separated from the other components by the floatation technique, all of the other components float on the water while the PET sinks.

3. The tensile strength of the milk bottle resin is high and what is expected for a high molecular weight blow molding resin of about 0.960 density. The recycle

detergent bottle resin is a lower density polymer for improved stress crack resistance and so is expected to have a lower tensile strength. The strength might also be affected by the presence of absorbed detergent that has not been completely removed during melt processing. The tensile strength of the base cup resin is low for several reasons. Two of the causes are: the low molecular weight of the HDPE resin used for the base cup, and as mentioned above, the presence of polypropylene which is incompatible with HDPE.

4. The elongations of the milk jug and detergent bottle resins are high as expected, but the base cup resin has a low elongation for the same two reasons as mentioned above.

5. The flex modulus of the milk jug resin is high and at the expected level for a high density HDPE. The detergent bottle resin has an expected lower modulus because of the lower density of the resin. The base cup resin, a 0.960 density resin, has a higher modulus than the milk bottle resin because of its viscosity allowing increased crystallinity and the presence of polypropylene from the cap and label.

6. The toughness of the resins was measured using the notched Izod impact test. The milk jug resin has the highest level of toughness of 19.0 - 21.0 ft-lbs/in.; the detergent bottle resin has a lower toughness, and the base cup resin is brittle as expected again because of its lower molecular weight and the presence of polypropylene.

Flow Improvement of HDPE Recycle

The HDPE recycle resin in most demand is that derived from milk bottles mainly because it is unpigmented. However as the resin is designed for blow molding it has a high molecular weight and therefore does not produce economical cycle times when injection molded. In order to define the problem further injection molding tests were run using a snake flow mold. The length of flow was measured for blends with a higher melt index injection molding resin. The results of these tests using a nominal 6 melt index 0.960 density resin are shown in Table III. (Conditions: 6-Ounce Injection Molding Machine, Melt Temperature: 220°C, 1.25 In Diameter GP Screw, 60 RPM, Fast Injection, and 20/20 Second Cycle.)

Table III
Snake Flow of Blends of Recycle HDPE Milk
Jug Resin and a Standard Injection Molding Resin

Resin Composition	Melt Index G/10 Min	Snake Flow at 700 PSI Inches	% of Control	Snake Flow at 1400 PSI Inches	% of Control
"Sclair" 2907	5.7	18.5	100	29.0	100
50/50 Recycle Milk Jug/"Sclair"	1.5	17.2	93	28.4	98
75/25 Recycle Milk Jug/"Sclair"	1.0	16.5	89	27.6	95
100% Recycle Milk Jug	0.6	15.4	83	26.0	90

The snake flow of two blends based on a 5.7 melt index molding resin and recycle milk jug resin were measured at two different injection pressures at a barrel and nozzle temperature of 220°C. The length of the flow of the compositions are compared to the two neat resins. It can be seen that the higher injection pressure of 1400 psi gives a greater increase in flow over the lower pressure for all three compositions containing the recycle resin.

It may be surprising that adding a resin having a melt index 1/10 of the control resin does not appear to reduce the flow as much as one would expect. But it must be remembered that the melt index test is run at a low shear rate and the shear rates used for this snake flow test are much higher and result in what is called shear thinning of the melt. This data also shows why some of the newest HDPE resins have a bimodal distribution of molecular weights which result in significantly improved flow over a resin of the same average molecular weight but with a narrow distribution.

Shrinkage of Moldings

We have found, however, that injection moldings produced from the blends containing the recycle HDPE milk bottle resin exhibit considerably greater shrinkage than do moldings from the neat HDPE injection molding resin. The amount of shrinkage increases as the amount of recycle resin increases. This effect is shown in Figure 1 which shows the amount of shrinkage after exposing a section of a side wall of a molded container to 160°C.

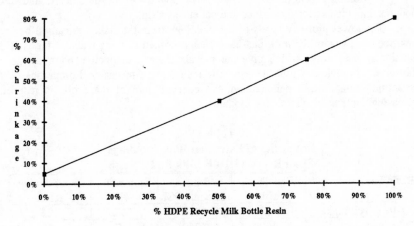

Figure 1. Effect of added recycle on shrinkage.

The amount of shrinkage varies linearly with the concentration of the high molecular weight fractional melt index HDPE recycle resin present in an 8 melt index HDPE injection molding resin. Blending 50% of a recycle higher flow HDPE base cup resin into the milk bottle resin, although reducing the shrinkage, does not bring the shrinkage back to the level of that of the neat molding resin. Depending on the configuration of the molding and the molding conditions, the dimensions of the molding may be significantly different for a blend with a high molecular weight recycle HDPE as compared to a control. This increased shrinkage of the molding is caused by the presence of higher molecular weight molecules which align themselves in the direction of flow during molding and therefore have an increased tendency to relax to a more random configuration.

Blow Molding Recycle HDPE

Since both the recycle milk jug resin and recycle detergent bottle HDPE resin came from containers that were blow molded, it is apparent that these two recycle resins should be readily blow molded. And that is the case. We have produced large extrusion blow moldings from the recycle HDPE milk jug resin and have evaluated the mechanical properties of the side wall of the moldings. Cylindrical moldings about 32 inches long and 8 inches wide were blow molded with wall thicknesses of 60 and 100 mils. Tensile and flex bars were cut out of the moldings and tested for mechanical properties with the results shown in Table IV.

The tensile strength, elongation, and flex modulus are shown for the two different molding thicknesses using the as received HDPE recycle flake and an extrusion-extracted and pelletized HDPE resin. The data indicate the following:

1. The 60 mil thick wall moldings have lower tensile strength and lower flex modulus than the 100 mil thick moldings.

2. Moldings from the as received flake have lower tensile strength and flex modulus than the extracted extruded and pelletized resin.

The lower property levels of the moldings from the flake are caused by the presence of oils from the milk that have been absorbed into the walls of the HDPE containers. These oils, which give the recycle flake a distinctive odor, act as a plasticizer as previously indicated, thereby reducing the mechanical properties. On extraction-extrusion and pelletization, the concentration of these oils is reduced increasing the tensile strength and modulus.

Table IV
Properties of Extrusion-Blow Moldings
From Recycle HDPE Milk Bottle Resin

Resin Feed	Wall Thick. Mils	Tensile Strength PSI	Elongation %	Flex Modulus KPSI
As-Received Flake	100	4010	547	125
Pelletized	100	4260	639	153
As-Received Flake	50	3860	440	87
Pelletized	60	4015	700	139

Melt Rheology of Recycle HDPE Resins

Resins used for extrusion blow molding have high molecular weight because they require high melt strength to prevent sagging of the extruded parison during the blow molding operation. As was pointed out previously, the recycle resins from both milk jugs and detergent bottles have absorbed some of the product in the containers which act like plasticizers reducing the polymer melt strength. However on pelletization of the recycle flake using an extractor extruder, the concentration of these absorbed materials is reduced increasing the polymer melt viscosity. The effect of the absorbants on the melt viscosity over a range of shear rates has been studied for both recycle blow molding resins.

One example of this difference in melt viscosity and shear stress for the recycle detergent bottle resin is shown in Figure 2 which plots these parameters over a shear rate range of 10 to 1000 reciprocal seconds at a melt temperature of 177⁰C. The significantly higher melt viscosity and sheer stress of the pelletized resin at low shear rates as compared to the as received recycle flake is to be noted. For equivalent blow molding performance, the as received recycle flake should be blow molded at a somewhat lower temperature than the pelletized resin. A similar but not as pronounced difference is observed for the recycle milk bottle resin.

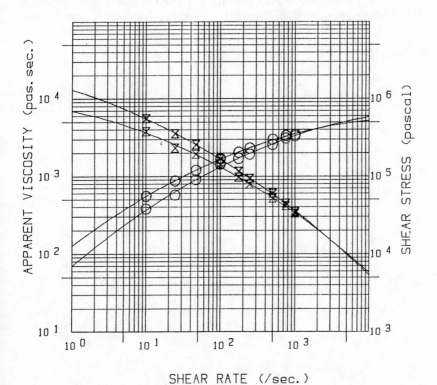

Figure 2. Effect of shear stress and viscosity
of recycled detergent bottle resin.

Recycle HDPE Blow Molded Barricade

Du Pont has an agreement with the Department of Transportation of the State of Illinois to supply them with items made out of recycle plastics for use on the highway. One of the items being developed is a traffic barricade produced from recycle HDPE milk bottle resin. This barricade, of a new design is about 38 inches high, 27 inches wide and about 3 inches deep and is readily extrusion blow molded from the recycle HDPE containing titanium dioxide pigment. About 50 of these signs have been installed on a highway in Illinois for about 6 months and found to be an effective replacement for the conventional barrels. With slight modifications to the design, this barricade is being commercialized.

Conclusions

High density polyethylene obtained from recycled post consumer containers can be readily fabricated into useful items by conventional injection molding and extrusion blow molding processes. The properties of moldings produced from the recycle material have properties very similar to virgin material. Commercial items are being produced and sold using the recycled HDPE. As the volume of recycled HDPE increases, the types of products made from these materials will increase and will be favorably received by the consumer.

RECEIVED May 18, 1992

Chapter 21

Multilayer Packaging in the Environment
One Company's Positions and Programs

S. J. Fritschel

Packaging & Industrial Polymers Division, Du Pont Polymers, P.O. Box 80011, Wilmington, DE 19880–0011

The solid waste issue has been one of the key topics of discussion in the packaging industry over the last two years. All packaging has been subjected to increasing scrutiny for its "friendliness" to the environment. Plastic packaging has been singled out by some as being a major contributor to the solid waste problem. Multilayer packaging, in particular, has been cited as being "unfriendly". This paper will try to separate myth from fact, and will discuss Du Pont's programs to address the issue of multilayer packaging from an environmental point of view.

The Scope of the Problem

Surveys have shown that the average person believes that well over half of what they throw away is plastic. In fact, the most recent studies by Franklin Associates show that plastics occupy about 20% of the municipal solid waste stream by volume. The same study shows that 40% of this stream is plastic packaging. In other words, 8% of the municipal solid waste stream by volume is plastic packaging.

The plastic packaging volume is broken down into 5.6% rigid packaging and 2.4% flexible packaging. This reflects the weight to volume advantage of flexible packaging over rigid packaging.

This paper is concerned with multilayer packaging, so these figures must be broken down further. The Flexible Packaging Association reports that 50% of all flexible packaging solid is coated or uncoated polyethylene. Another 33% is oriented polypropylene or other polypropylene monofilms, 13% are coextruded and 4% are laminated films. If one counts all coextruded and laminated films as multilayer, then the volume contribution to the solid waste stream of flexible multilayer films is 17% of 2.4%, or 0.4%. For rigid

0097–6156/92/0513–0266$06.00/0

packaging, the Plastic Bottle Institute reports that 90% of all bottles are mono-material bottles. Figures are very similar for tubs, trays, and other rigid, non-bottle packages. So the volume contribution for rigid multilayer packaging is, at most, 10% of 5.6% or 0.6%.

This means that, in total, multilayer plastic packaging comprises, at most, 1% by volume of the municipal solid waste stream.

Despite this relatively small contribution to the problem, multilayer packaging has been a target of controversy in the solid waste debate. Several proposals to ban multi-layer plastic packaging were made during 1989 and 1990 legislative sessions in several areas of the country. These proposals were justified based upon the perceived lack of recyclability of the multilayer packages. Du Pont and others have developed the data which refute these claims.

The integrated waste management system recommended by the EPA for the management of the municipal solid waste stream also applies to multilayer plastic packaging. That system includes source reduction, recycling, incineration, and land-filling.

Source Reduction

In many instances, multilayer packaging is the most efficient way to deliver the required shelf life for a product. Use of a thin layer of a high barrier polymer like ethylene vinyl alcohol (EVOH) provides product protection that would require the use of far more of a mono-layer polymer. In many cases, multilayer plastic packages replace more traditional glass and metal packaging. The replacement of #10 metal cans by flexible multilayer pouches reduces both the weight and volume of waste generated by at least 60%. Weight savings of multilayer plastic packaging over metal and glass also reduces the amount of energy used in product distribution as well as reducing waste due to breakage or dented product.

More efficient barrier, adhesive, and sealant resins for multilayer packaging are being developed. For example, a new EVOH type resin offers three times the barrier per mil of traditional EVOH resins. This can allow the packager to either increase shelf life at a constant thickness of barrier resin or to downgauge the barrier resins while maintaining shelf life. Another material in the development stage will offer increased flexibility to re-use process regrind, thereby reducing the amount of scrap generated by the converter.

Recycling

Unlike packaging made of homogeneous plastics like HDPE milk, juice, and water jugs or readily separable plastics like PET soda bottles, multilayer packages contain several different polymers and adhesives that cannot easily be separated into their component parts for recycling. This has led to the erroneous view that multilayer plastics cannot be recycled. A number of companies have shown that it is not necessary to separate multilayer packages into single components to recycle them.

Because it is probably the best known use of multilayer packaging, the ketchup bottle has been singled out as *the* example of non-recyclable packaging. A joint program with the Technical Committee of the Plastic Bottle Institute of the SPI helped demonstrate the recyclability of the multilayer PP/EVOH ketchup bottle.

In order to replicate real recycling conditions as closely as possible, the bottles used for this project were bottles manufactured for commercial use by American National Can, Continental Can, and Owens-Brockway. The bottles were filled with ketchup by Del Monte, Heinz, and Beatrice/Hunt-Wesson. The containers were emptied and then shipped to the Center for Plastics Recycling Research at Rutgers University.

The containers that arrived at Rutgers for processing were rinsed to remove any residual ketchup. The bottles went through the same process as that used for PET bottles:

- granulated to reduce to flakes
- put through a cyclone with a baghouse to remove paper remnants
- washed with soap to remove dirt and traces of ketchup
- rinsed to remove traces of soap
- put through a hydrocyclone separator (used for post-consumer soft drink bottles to separate PET from high-density polyethylene used for base cups)
- passed through a dryer, which also removes fine particle contamination
- packaged in gaylord boxes for shipping.

In this test project, bottles were processed with caps on. Ketchup bottle caps are made of polypropylene, the same resin as the bottle, but the caps are pigmented; in addition, they have an aluminum foil liner. Asking consumers to remove bottle caps (as they do with glass bottles) before recycling the containers will result in a clearer recycled flake; however, if caps are included during processing, the material will still perform like polypropylene but will have the pigment color.

The flake received from Rutgers was used to formulate a thermoplastic olefin (TPO). The final product contained 75% ketchup bottle flake with the other 25% composed of EPDM rubber tougheners and anhydride modified EPDM compatibilizers. TPO compositions based on virgin polypropylene would contain similar amounts of compatibilizers and tougheners. The small amount of aluminum present from the package inner seals was removed by melt filtration during the compounding step.

The compounded material, in pelletized form, was shipped to Du Pont's Troy, Michigan Automotive Development Center for molding trials. A 4000 ton Cincinnati Milacron injection press was utilized, together with a production mold for a 1984 Chevrolet Cavalier rear bumper fascia. A 4.75 pound shot was molded on a 99 second cycle. Trimmed part weight was 4.45 pounds. Injection molding pressures were 1200 psi injection, 900 psi pack and 750 psi hold. Back pressure was maintained at 50 psi, and clamp pressure was 3300 tons.

The compounded material processed satisfactorily and no problems were encountered during molding. The parts produced were yellow due to the incorporation of yellow caps during the recycling process.

The composition was tested by standard ASTM methods and compared with a commercial TPO based on virgin polypropylene. The results are shown in Table I.

Table I. Comparative Property Data

Property	Temp.	Commercial TPO Genesis AP 8210*	Ketchup Bottle Composition
Tensile Strength	R.T.	2,200 psi	3,600 psi
Flexural Modulus	R.T.	90,000 psi	75,400 psi
Gardner Impact	R.T.	>320 in-lbs	>320 in-lbs
	-30°C	>320 in-lbs	>320 in-lbs

*data from product literature

The experimental composition provided a reasonable match for the commercial composition. If additional flexural modulus were needed, the ratio of rubber to bottle flake could probably be modified to produce a more exact match without losing the critical impact properties.

Injection molded plaques of the experimental TPO were painted at Du Pont Automotive Products' Troy Laboratory. A variety of paint systems were evaluated versus steel controls. Of the systems tested, the most satisfactory was a coat of Du Pont 800R adhesion promoter followed by a coat of Du Pont Centari 871/2 base coat and then a layer of Centari RK3939 flexible clear coat. This system provided results for adhesion, cold crack, gloss, and distortion of image which compared favorably with systems used commercially for automotive components. (Table II) This system was used to paint the molded fascia.

Table II. Painting Study Results

Substrate	Experimental TPO	Steel Control
Adhesion Promoter	800R	None
Primer	None	764-189GX
Base Coat	872-Black	872-Black
Clear Coat	RK-3939	RK-3939
Initial Gloss	89	90
Initial Distortion of Image	81	78
Initial Adhesion (samples/passes)	10/8	10/10

Humidity Test X Tape/Cross Hatch

96 Hr.

Adhesion (samples/passes)	10/10	10/10
Blistering	None	None
Dulling	None	None

240 Hr.

Adhesion (samples/passes)	10/10	10/10
Blistering	None	None
Dulling	None	None

504 Hr.

Adhesion (samples/passes)	10/9	10/10
Blistering	None	None
Dulling	None	None

This test showed that multilayer PP/EVOH bottles could be handled using existing recycling processes and that the recycled material could be substituted for virgin polymer in a demanding end use.

Incineration

Recycling of multilayer plastic packaging may not always be viable due to collection, separation, or other economic issues. In these cases, waste-to-energy incineration is the preferred method of disposal. Like all plastic materials, multilayer plastics are an efficient contributer to the clean burning of municipal solid waste. Their high BTU content makes them especially desirable in waste-to-energy systems. Data from the EPA and other sources suggest that properly designed and operated waste-to-energy facilities can recover 40% of the energy stored in the plastic, while still meeting all environmental regulations.

Du Pont is working actively to help ensure that state of the art waste-to-energy incineration is built in various parts of the country. For example, in Brevard, North Carolina and Chattanooga, Tennessee, steam and electricity generated from municipal waste-to-energy incinerators are purchased for use in manufacturing operations. These long term purchase contracts allow the incinerator operators to obtain financing at attractive rates.

At Parkersburg, West Virginia, the company also operates its own plastic waste-to-energy incinerator. This incinerator handles non-recyclable, non-hazardous waste generated in our own processes, and is expected to also accept similar materials from our customers. The facility is capable of handling six thousand pounds of waste per hour which would generate 27,000 pounds per hour of steam. The steam is used in polymer production operations at the site. Construction of this facility allowed the plant to defer construction of new coal-fired boilers.

Landfill

Properly designed landfills are an essential part of any integrated solid waste management system. If the more desirable recycling or waste-to-energy incineration facilities are not locally available, multilayer plastic packaging can be landfilled.

Acknowledgments

The cooperation of the Technical Committee of the Plastic Bottle Institute and the technical contributions of Dr. P. M. Subramanian are gratefully acknowledged.

RECEIVED June 1, 1992

Chapter 22

Recycling of Reaction Injection Molded Polyurethanes

William J. Farrissey[1], Roy E. Morgan[1], Rick L. Tabor[2],
and Melissa Zawisza[2]

Polyurethanes Department, The Dow Chemical Company, [1]Building
B–1608, [2]Building B–1470, Highway 288, Freeport, TX 77566

Several processes have been developed to recycle RIM thermoset
polymers, including reuse of reground scrap, compression
molding and extrusion, recovery and recycle of chemical
components, and energy recovery. For recovery and recycle of
scrap from Reaction Injection Molding (RIM) processing, size
reduction is a critical first operation. RIM scrap is granulated in
a first step, and subsequently pulverized to 180-300 micron size
in an impact disc mill. At this size, the RIM regrind can be used
as an inexpensive extender in the original applications, such as
automotive fascia, with no loss in physical or mechanical
properties. Surface quality of the parts is excellent. RIM regrind
can also be used as a filler in thermoplastics such as
polyethylene. Compatibilizing polymers have been developed
which allow incorporation of 25% polyurethane RIM regrind into
polyethylene without appreciable loss of impact properties.
Energy recovery from polyurethane RIM scrap can be
accomplished in suitable combustors, without generating
excessive stack emissions. These recycle/recovery technologies
augment the favorable energy savings which accrue from the use
of lightweight polyurethane parts in automotive applications.

Polyurethane plastics constitute the largest, and by far the most diverse, of the
thermoset resins. Over 3 billion pounds of polyurethane materials were sold in
the United States in 1990(1). Flexible foams for bedding and furniture
applications consumes the largest volume, about 1.7 billion pounds; rigid foam,
about 875 million pounds; and elastomer applications, about 650 million
pounds, of which Reaction Injection Molding (RIM) represents 212 million
pounds.

0097–6156/92/0513–0272$07.00/0
© 1992 American Chemical Society

Polyurethanes are formed by the controlled reaction of multifunctional isocyanates with multifunctional polyols. If blowing agents are present during the polymerization process, a foamed product results. The foamed products have a variety of production processes and applications. Examples include pour-in-place rigid urethane foam as insulation in refrigerators, foamed continuous blocks of flexible urethane foams which are then cut into shapes or, as in the case of auto seats, the foam may be molded into its final shape. Elastomeric polymers such as automotive exterior parts are molded into their final form *via* the RIM process.

Scrap is generated in these various production operations. Virtually all of the scrap from the flexible foam production and cutting of the bunstock into various shapes is recovered and reused as carpet underlay. In this process, the foam scraps are first chopped, then coated with an isocyanate prepolymer binder. The coated foam is then compressed and the binder cured with steam. The blocks of rebonded foam are then cut into the required thicknesses. In this manner, about 280 million pounds of domestic scrap and 120 million pounds of imported scrap, otherwise destined for landfill are converted into useful product(2).Nissan has described its technology to recover polyurethane foam and the associated polyvinyl chloride fabric(3). Both components can be recycled into automotive parts. Air Products has described a technique to reuse scrap flexible foam in the foam production process. About 10% of cryogenically ground foam scrap can be incorporated into new foam production(4).

Another recycling technology, which is receiving renewed attention of late, is chemical recycling - conversion of polymer waste into monomer chemicals. This process was examined extensively by General Motors(5) and Ford Motor Company(6), who explored hydrolysis methods to recover chemical values from auto seating scrap. The Upjohn Company developed glycolysis methods to convert polyurethane scrap from a variety of sources into polyols which could be reused in rigid foams, especially(7). It has been reported that this process has been practiced on an industrial scale.

Recently, glycolysis development has proceeded with Prof. Bauer(8) at the Technology Institute, Aalen, and with Prof. Simioni at the University of Padua(9). Prof. Bauer has demonstrated recovery of polyols from car seats and attached polyester or nylon fabric, and Prof. Simioni has recovered polyols from microcellular elastomers such as shoe sole scrap.

These technologies have been applied to scrap from RIM operations with encouraging results. Table I. shows the physical property comparison of a RIM elastomer with similar material made by substituting 10% and 20% of the original polyol with polyol recovered by glycolytic cleavage of RIM scrap(10). Clearly the comparison is quite heartening. Most importantly, impact strength, both at ambient temperature and at -25°C, were essentially unchanged by the incorporation of recycle polyol, for both painted and unpainted specimens.

Table l. RIM Properties With Recycle Glycolysis Polyol

	Original[a] +10% GLYC	Original[a] +20% GLYC
Specific Gravity	99	99
Tensile Strength,	100	108
Elongation, (%)	107	125
Flexural Modulus,	82	113
Hardness, Shore D	100	105
Impact Strength,		
Unpainted, 23°C	N.B.	N.B.
Unpainted, -25°C	100	103
Painted, 23°C	N.B.	N.B.
Painted, -25°C	99	102
Water Absorption, (%)		
140 hrs.	117	92
Shrinkage, (%)	94	100

[a]Data expressed as *per cent* of original properties.

Until recently, RIM polyurethane and polyurea polymers have been perceived as being unrecyclable. Current recycle efforts have shown that RIM scrap materials can be recycled *via* several technologies. One technique, described recently by Mobay, involves compression molding of the ground scrap(*11*). The scrap, reduced to a granular size is preheated to 175°C, placed in a mold at 175°C, compressed while hot at 200-400 bars (2800-2600 PSI) for 3 min., and demolded. Cycle times depend on the oven heating capabilities, molding temperature and pressure. Compression molding has continually improved as the process becomes closer to a commercial reality. The granulated polymer has been used in two size ranges: 1-5mm and 0.2-1 mm. The best results have been achieved with the 0.2-1mm size. Parts have been produced with flow paths of 25-30 cm and the in-mold times have been reduced to 60 seconds (615 bar). The mechanical properties are shown in Table ll. In this example the thermal properties of the material as measured by the heat sag test were improved. This is due to the improved phase segregation resulting from the elevated processing temperatures of the compression molding process. The properties, while not quite equal to the original are adequate for some non-appearance parts, such as sight shields and deflectors. Costs are claimed lower than injection molding of typical engineering thermoplastics. Extruded profiles from RIM scrap have also been reported by ICI(*12*). Both of these processes utilize 100% RIM scrap material.

Table II. Compression Molding

Mechanical Property Comparison
(20% MGF Reinforced Polymer)

	Original	Recycled
Specific Gravity	1.21	1.22
Flexural Modulus, MPa	1200	1050
Tensile Strength, MPa	25	15
Elongation, %	140	40
Heat Sag, mm	14	5
(160°C, 100 mm, 60 min.)		

Another process, described here, incorporates ground and pulverized RIM scrap as an extender or non-reinforcing filler in virgin RIM chemicals. RIM molders typically generate 5-10% scrap consisting of the sprues and runners from each part, molding defects and paint defects. It was decided to develop a simple process by which the molder could reuse this scrap to improve the economics of molding today's RIM fascia, trim and body panels as well as other plastic molding operations. The result of these efforts was a process of reducing the size of the thermoset polymer to a very fine powder for use as an extender in both virgin thermoset and thermoplastic polymers.

Size Reduction Process

Until recently, pulverization of scrap RIM polymers was not an economical process because it required the use of cryogenic grinding which added a prohibitive cost to the size reduction process. However, in 1987, a technique was discovered to economically grind the RIM thermoset scrap into a size useful as a filler in other applications. This technique uses impact disc mills which employ air during the pulverization to keep the polymer cool and to avoid agglomeration and polymer degradation. The size reduction process is not a cryogenic process. Runners, sprues and scrap parts resulting from molding and paint defects are gathered, are granulated, and finally pulverized in an impact disk mill to 180-300 microns (-80 to -50 mesh).

A process schematic as shown in Figure 1 *(13)* depicts the major steps in the size reduction process.

The surge bin/silo is added to compensate for the different grinding rates of the granulator and pulverizer. The granulators are capable of processing the polymers at rates of several hundred pounds per hour while the impact pulverizer that we have worked with operates at a 100-300 lb/hr. rate. One granulator can thus accommodate 3-4 pulverizers. Figure 2 illustrates the resulting particle size from the granulation and pulverization steps *(13)*.

Three manufacturers of impact disc mills have been identified that have equipment capable of achieving the desired particle size reduction. These companies are Wedco, Inc.*(14)*, Herbold Granulators U.S.A., Inc.*(15)*, and Pallmann Pulverizers Company, Inc.*(16)* and their European associates.

A variety of RIM polymers has been successfully pulverized on this equipment, including painted, unpainted, filled and unfilled RIM fascia, body panels (fenders) and side claddings. As expected, the unfilled fascia polymers, because of their higher elastomeric character, exhibited the lowest pulverization efficiency. As polymer rigidity is increased through the addition of fillers, or as the polymers are painted, the pulverizing efficiency is increased (Table lll) *(13)*.

Table lll. Scrap Source *VS*. Pulverizing Rates

WEDCO SE-12-TC
-30 mesh sieve

Polymer Type	Pulverizing Rate, lbs/hr
Unfilled	245
10% Milled Glass	290
15% Milled Glass	415
10% Glass, Painted	380
15% Glassed, Painted	390
Unfilled, Painted	320

SOURCE: Reprinted with permission from ref. 13. Copyright 1991 Society of Automotive Engineers, Inc.

Figure 3 depicts the effect of final particle size on the grinding rate which is lowered as the particle size is reduced. Work has been conducted with particle sizes of 300 microns (-50 mesh) and smaller. The preferred particle size is in the range of 200 microns (-80 mesh) and smaller to minimize the effect on part surface quality. Figure 4 illustrates the particle size distribution of a typical -80 mesh RIM regrind *(13)*.

RIM Regrind as a Filler

After the size reduction process, the material can be handled in exactly the same fashion as other fillers. Development work has incorporated regrind into a variety of other applications such as: low density mat molded and low density RRIM composites, energy management foams, numerous thermoplastics, RIM and RRIM fascia, and RRIM body panels and side claddings. RIM molders now

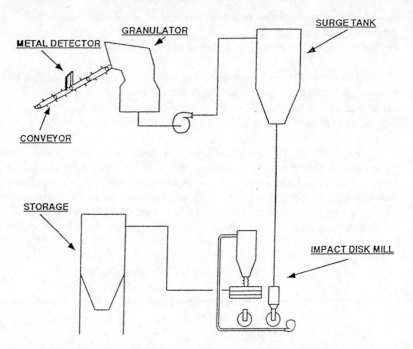

Figure 1. Size Reduction Process (Reproduced with permission from ref. 13. Copyright 1991 Society of Automotive Engineers, Inc.)

Figure 2. Particle Sizes (Reproduced with permission from ref. 13. Copyright 1991 Society of Automotive Engineers, Inc.)

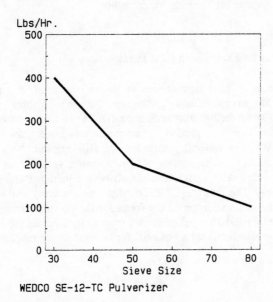

WEDCO SE-12-TC Pulverizer

Figure 3. Effect of Particle Size on Grinding Rate (Reproduced with permission from ref 13. Copyright 1991 Society of Automotive Engineers, Inc.)

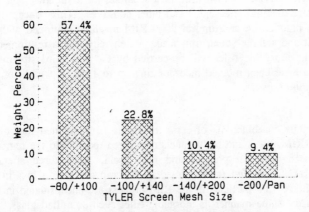

Figure 4. -80 Mesh Particle Size Distribution

have a large variety of options to choose from in order to match their recycle needs to their manufacturing capabilities. Since most RIM molding operations utilize some type of inorganic filler, working with and handling the regrind should present no unusual handling requirements.

RIM Regrind as an Extender in RIM Parts

The incorporation of RIM regrind into fascia will be used to illustrate this technology. This process technology requires the use of a three-stream RIM machine when using diethyl toluenediamine (DETDA) as the chain extender. The three-stream process is needed to prevent an unwanted side effect where DETDA (a strong polar solvent) adsorbs into the RIM regrind. This adsorption prior to impingement mixing results in stoichiometric variation of isocyanate and amine, an index drift, which leads to off-ratio conditions and subsequent property changes. The unreacted DETDA that was absorbed in the regrind is then free to diffuse through the newly formed RIM polymer, causing surface discoloration before and after painting. Attempts to incorporate the regrind into the isocyanate stream indicated a potential for unwanted side reactions with the isocyanate.

Three Stream RIM Processing

The process schematic in Figure 5 shows the flow of the materials *(13)*. Regrind is added to the polyol stream, the isocyanate is the traditional second stream, while DETDA is dispensed in a separate, third stream along with the catalyst, internal mold release*(17)* and other additives. This concept has been reduced to practice on an Admiral 9000 RIM machine equipped with a third stream unit and a three stream mixhead. When recycling a reinforced RRIM polymer, the inorganic filler will be carried by the new third stream. Large front fascia have been molded incorporating up to 10% by weight of the low density regrind powder.

Various combinations of scrap materials from painted and unpainted RIM and RRIM fascia and body panels have been reduced to the regrind form and processed into test plaques and full fascia. The three stream RIM processing of the regrind from painted RIM scrap creates no loss in surface quality. The regrind particles in the RIM process are encapsulated in the new polymer in the same manner as inorganic filler such as milled glass. Typical RIM processing conditions are listed in Table IV. The development work has centered around a level of 10% regrind (by weight) *(13)*. Recent work on the three stream process has used loadings of up to 15% RIM regrind.

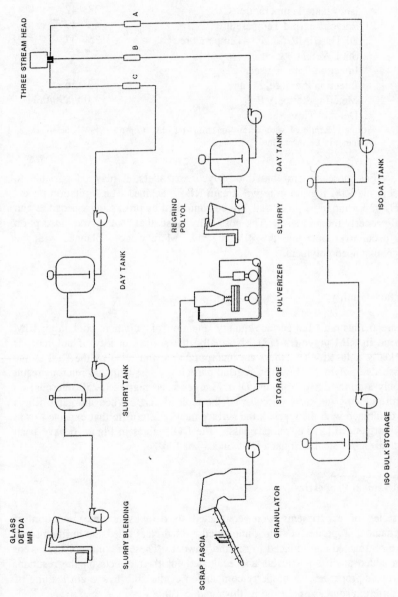

Figure 5. Three Stream RIM Process (Reproduced with permission from ref. 13. Copyright 1991 Society of Automotive Engineers, Inc.)

Table IV. Processing Conditions

Molding Conditions
Production Scale Front Fascia

A: Isocyanate Temperature °C	38 - 42
B: Polyol/Regrind Temperature °C	46 - 49
C: DETDA/IMR/Catalyst Temperature °C	35 - 37
Shot Weight, kg	5.7
Injection Rate, Kg/sec.	5.4
Injection Pressure, MPa	13 - 16
Metering Ratio, A:B:C	1.00/0.90/0.34
Demold Time, sec.	30

SOURCE: Reprinted with permission from ref. 13. Copyright 1991 Society of Automotive Engineers, Inc.

For those systems based on glycol extenders, it may be possible to process the RIM regrind in a two stream RIM machine. The limitation on the levels of regrind to be processed will be controlled by processing viscosities and final property requirements. The possibility exists that some glycol adsorption will occur over time into the RIM regrind. More work is planned with the glycol extended polymers.

Surface Quality

Figure 6 illustrates the surface quality retention of polymers containing RIM regrind in RIM polymers *(13)*. Normally, the surface quality of polymers is markedly reduced when fillers are incorporated as indicated by the DOI of the glass filled polymer. However, regrind particles of less than 200 microns result in only a minor decrease in the DOI. Larger sized regrind (>250 microns) do begin reducing the surface quality of the molded part. When inorganic fillers are combined with the regrind, the surface quality returns to that expected from the inorganic filler, i.e. milled glass. The DOI values in Figure 6 have been normalized for steel to equal a DOI "index" of 100.

Mechanical Properties

A series of experiments were conducted to demonstrate the retention of mechanical properties. Painted, unpainted, milled glass filled and unfilled scrap were segregated and reduced to regrind powder. These regrind powders were then added at different levels and evaluated for their effects on the resultant polymer's properties. A property comparison of the 10% by weight loading of the different sources of regrind is illustrated in Table V.

Table V. Mechanical Properties at 10% by Weight Loading

10% Regrind, -80 mesh
Source/Type of Scrap

	No Regrind	Unfilled, Painted	Filled, Painted
Specific Gravity	1.04	1.05	1.08
Flexural Modulus, MPa	350	328	365
Tensile, MPa	24.9	26.0	24.0
Elongation, %	240	265	260
Tear Strength, kN/M	100	100	101
Heat Sag, mm - 6", 1 hr/121°C	18	18	18
Gardner Impact, J			
24°C	27.1	27.1	25.3
-29°C	>36.2	>36.2	>36.2

SOURCE: Reprinted with permission from ref. 13. Copyright 1991 Society of Automotive Engineers, Inc.

The "regrind" filler is non-reinforcing since it has a similar flexural modulus to that of the matrix resin. Hence there is no significant effect on physical properties in the new composite polymer. One can visualize RIM regrind as a low cost bulking agent. As evidenced by examining Table V, the incorporation of up to 10% of the regrind has no appreciable effect on the polymer flexural modulus, heat sag (thermal stability) or toughness as measured by tensile, elongation and low temperature impact properties. What is even more important is that the recycle polymers pass the specifications for virgin polymers.

Regrind in Filled Systems

Built upon these encouraging results, three-stream processing trials were conducted in Europe which further proved the capability of this technology. These trials were conducted at Afros Cannon in Italy and Krauss Maffei in Germany. Each of the RIM streams were filler capable. This allowed the third stream to carry a typical inorganic filler. Molded polymers were produced that contained 10-15% RIM regrind (from a mixture of scrap parts) and 15-17.5% milled glass fiber. The DETDA containing stream contained the milled glass fiber while the polyol stream carried the RIM regrind.

As the regrind composition incorporates polymers with much higher glass loadings, such as glass-mat containing polyurethane (SRIM), there is a reinforcing contribution from the regrind. This can be seen in the polymer containing 15% regrind in Table VI. The glass mat in the SRIM polymer has a different surface treatment and its effect can be seen in the reduction of the elongation. The regrind containing RRIM polymers also meet the specification requirements.

Table VI. Properties of Regrind in Glass Filled Polymer (-80 mesh)

		Standard System	Standard + <200 µm		
Polyol		XZ95403.28	XZ95403.36 XZ95403.37		
Isocyanate		M342	M342		
Filler	DIN Standard	15%MF7901	15%MF7901 10% Regrind	17.5% MF7901 10.0% Regrind	17.5% MF7901 15.0% Regrind
Flexural Modulus, N/mm^2	53457	900	900	1259	1253
Tensile Strength, N/mm^2	53455	28.0	25.6	26.7	25.6
Tensile Modulus, N/mm^2	53457	750	673	841	913
Elongation, %	53455	160	149	137	85
Hardness, Shore D	53505	59	63	65	65
Density, gm/cc	53479	1.18	1.20	1.21	1.21
Post Cured, min/°C		45/120	45/120	45/120	45/120
Index, NCO/H$^+$		110	110	110	110

Equipment Capital and Economics

Recycling, in addition to the beneficial environmental effects of the regrind process, should also reduce the manufacturing costs to the molder. The cost of molding a typical RIM fascia can be reduced by incorporating the RIM regrind (Figure 7).

Economics were determined by considering the direct capital, tooling, labor energy and raw material costs of the RIM process to mold a part(*18*). The required incremental capital investment to process RIM regrind consists of size processing equipment located at the molding site along with filler capable third stream processing equipment. Size reduction capital is on the order of $250,000 which includes a granulator, two pulverizers, silos and conveying equipment. Additional blending and batch tank capital will be required to handle the new third stream component ($850,000 used in the model). The third stream injection unit capital is about $400,000 which includes a filler capable injection cylinder, tanks, hydraulics and two three stream mixheads. The third stream

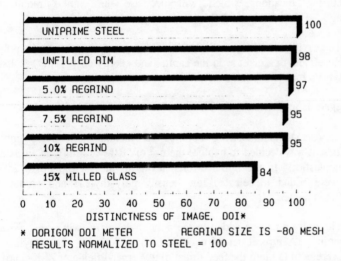

Figure 6. Painted Surface Quality (Reproduced with permission from ref. 13. Copyright 1991 Society of Automotive Engineers, Inc.)

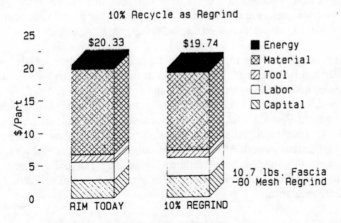

Figure 7. Molded Part Cost

unit can feed more than one RIM press.   In the cost model, the third stream capital cost is spread over two presses.   The size reduction and blending capital cost is spread over all the clamps of a molding plant assumed to operate 12 presses.   Many RIM molders already have portions of this equipment at their manufacturing   locations   which   will   decrease   the   capital   requirements accordingly.   The capital was depreciated over a five year period.

The savings in purchased raw materials by recycling the scrap polymer more than offsets the increase in the capital and energy cost.   This results in a rapid payback for the capital investment.

RIM Regrinds as Extender in Thermoplastics

Since many RIM manufacturing facilities also include thermoplastic molding capabilities, it was decided to explore the use of "RIM regrind" as an extender in thermoplastics.   In the initial efforts, levels of the regrind up to 25% by weight were added to selected low density polyethylene and compression molded into plaques.

The regrind can be added to the polyethylene *via* twin screw extrusion compounding.   A typical set of conditions is as follows.   Polyethylene and regrind were both fed into the feed throat of a Werner-Pfleiderer ZSK-53/5L co-rotating, twin screw extruder operating under the following barrel temperatures: Zone 1 = 180°C, Zone 2 = 185°C, Zone 3 = 190°C, Zone 4 = 190°C.   The regrind was metered using a Plasticord screw powder feeder.   A vacuum of 29 inches of mercury was maintained on zones 3 and 4 to devolatilize any water which might have absorbed into the regrind powder.   In addition to the use of the mass balance, precise analysis can be accomplished *via* Soxhlet extraction of the polyethylene from the thermoset regrind composite using toluene.

The addition of up to about 10% regrind did not significantly effect tensile, tear and flexural modulus properties of these polyolefins.   As the level was increased above 10% upwards to 25%, the properties were reduced.   This is due to insufficient bonding of the polyolefin to the RIM regrind at higher loading levels.   Poor interfacial adhesion causes microscopic fracture sites which yield a loss in properties.   Examination of a cold fracture surface of a conventional polyethylene/RIM regrind composite (25% regrind, -80 mesh) by SEM supports this mechanism of failure (Figure 8).   The polyethylene was pulled out of the cell structure of the nucleated RIM polymer, leaving the nodule-like appearance.

Work was conducted to design a modified polyethylene copolymer which would achieve improved bonding at the polyolefin/polyurethane filler interface.   This improved interfacial adhesion is illustrated in Figure 9 which is a SEM photomicrograph of a cold fracture surface of a modified polyethylene copolymer/RIM regrind composite (25% regrind).

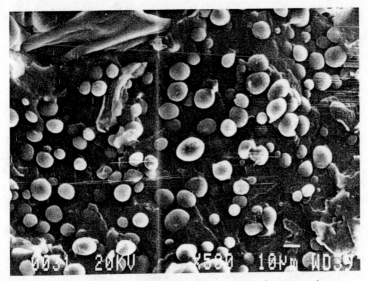

Composite of 25% RIM Regrind (-80Mesh)
with 75% ATTANE* 4001 ULLDPE
(200X Magnification)

Figure 8. Cold Fracture Surface

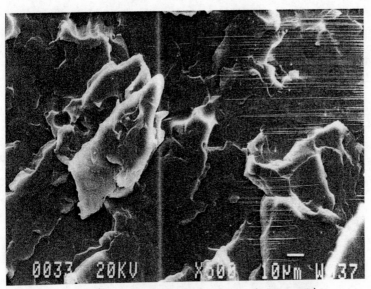

Composite of 25% RIM Recycle (-80 Mesh)
with 75% Modified ATTANE* 4001 ULLDPE
(200X Magnification)

Figure 9. Cold Fracture Surface of Modified Polyolefin

The improved adhesion obtained in the modified polyolefin translates to improved physical properties in the composite such as elongation and notched Izod. Table VII illustrates the data that were generated with the DOWLEX* 2035 LLDPE/RIM regrind composites.

Table VII. DOWLEX* LLDPE 2035/RIM Regrind Composites (-80 mesh)

	No Regrind	10% Regrind	25% Regrind	Modified Polyolefin 25% Regrind
Specific Gravity	0.912	0.923	0.952	0.957
Tensile Modulus, MPa	105.1	111.5	86.7	135.8
Tensile Strength, MPa	10.6	9.2	7.7	12.1
Elongation, %	481	452	87	427
Flexural Modulus, MPa	263.4	316.5	336.5	330.3
Tear Strength, kN/m	116.2	92.9	61.6	99.9

Similar results were seen with the modified ATTANE* LLDPE. Figure 10 shows the effect of RIM regrind on notched Izod impact at a level of 25 weight percent.

An 8% RIM regrind/ATTANE LLDPE composite was injection molded using the conditions listed in Table VIII. While these conditions were not optimized, the material exhibited no differences in processability from the virgin ATTANE LLDPE.

Table VIII. Injection Molding Conditions

	8% Regrind, <200 µm 92% ATTANE LLDPE
Barrel Temperature, °C	170
Mold Temperature, °C	27
Injection Pressure, Bars	40
Cycle Time, Sec.	30

*Trademark of The Dow Chemical Company

The rheology of the modified polyolefin and the 8% regrind composite shows an apparent viscosity increase (Figure 11) at low shear rates. Further studies are being conducted to examine other aspects of thermoplastic part manufacture using these RIM regrind containing thermoplastics.

The compounding of regrind into polyethylene incurs a cost based on the amount of thermoplastic being compounded. To maximize the economic benefits of this recycle route, a regrind "concentrate" containing 40-50% regrind in polyethylene can be made by compounding the regrind and the polyethylene at the molding site. This involves the expenditure of capital for compounding equipment. Alternately, scrap thermoset parts or powder can be shipped to a custom compounder along with unmodified polyethylene which may then be made into a concentrate. This concentrate would then be shipped back to the molder for dry blending with virgin pellets to the desired level of recycle in the final injection molded part. Savings are hence enjoyed *via* the reduced use of the thermoplastic raw materials as well as reduced landfill costs.

In summary, this work illustrates that RIM regrind from scrap RIM polymers can effectively be incorporated into various polyolefin polymers at loadings up to 25% by weight. Application examples could include fender skirt liners, sight shields, trunk/boot liners, blow molded under-the-hood fluid containers, profiles, extruded interior or exterior trim, etc.

Energy Recovery

In common with most plastics, polyurethanes have an appreciable energy content, 28-31 MJ/kg (12,000-14,000 BTU/lb), some portion of which is recoverable in an appropriate combustion facility. A study of the combustion of RIM materials in several types of industrial facilities was conducted jointly by the Dow Chemical Company and Mobay Corporation(19). Two of the units were equipped with energy recovery facilities. The Dow unit is a rotary kiln with an after burner, operating at 870-1200°C to ensure 99.99% destruction of specific organic constituents. Gases from the incinerator enter a multipass boiler to recover steam values, then enter a quench chamber and scrubber to initiate acid gas and particulate removal. The Mobay fluidized bed combustor operates at 900°C and contains about 12,700 kg of silica sand. The churning of the sand under the influence of the combustion air provides very high combustion efficiency. Again, heat is recovered in a 3-stage boiler. The cooled gases pass through an electrostatic precipitator for removal of particulate, and through a 2-stage scrubber to remove acid gases.

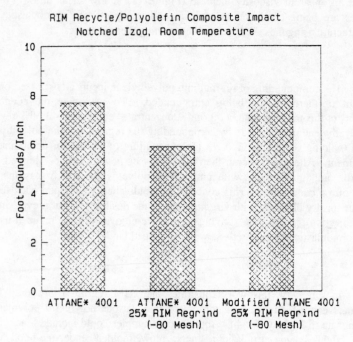

Figure 10. Notched Izod Impact Data

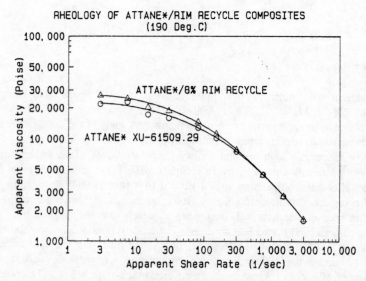

Figure 11. Rheology Curves

The combustion data are summarized in Table IX. Combustion rates were 135 kg/hr (300 lb/hr) and 500-550 kg/hr (1100-1200 lb/hr). Oxygen values for the fluidized bed were 7% and for CO, 80-100 ppm. (The values for the kiln were 7-12% and 0-2 ppm respectively.) NOx values were low for both technologies; 60±10 ppm compared to a base line 55±10 ppm for the fluidized bed, for example. The combined emissions for RIM combustion - CO,NOx, and particulate were compared to coal, oil and natural gas for the production of the same quantity of steam (Figure 12). Only the natural gas was lower than RIM for total emissions. Since RIM contains no sulfur, no SO2 is produced in its combustion, unlike coal and fuel oil. Hence, thermal recycling with energy recovery is feasible; fuel gas emissions are low. Potential uses include as a fuel in municipal sludge incinerators, waste-to-energy facilities and in-plant steam production.

Table IX. RIM Scrap Test Burns Operating Conditions and Results

	Mobay Fluidized Bed	Mobay Mass Burn	Dow Rotary Kiln
Feed Rates lbs/hr	250/1200	300	300/1100
Temperature (°C)	900	930	920
Oxygen %	7	3.5-13.7	7.5-12
CO ppm	50-100	55-175	0-2
Steam Rate (lb/hr)	15,000	--	60,000
NOx ppm			
Baseline (Natural Gas)	55[1]	70[2]	38[3]
RIM Enriched Feed	60-200[1]	67-186[2]	47-75[3]

SOURCE: Adapted from ref. 20.

Life Cycle Analysis

It is useful to examine the consequences of these recycle activities on the life cycle energy requirements for automotive parts. If we look at the energy expenditures to make a steel part - that is the energy required for mining, smelting, casting of ingots, rolling of sheet and stamping of parts, we find the 44.5 MJ of energy are required per kg of steel (19.200 BTU/bl) Table X(20). For polyurethane parts, the energy for petroleum production and refining, conversion to chemicals and processing into parts is 73.5 MJ/kg (31,700 BTU/lb). On the same scale, gasoline is 54 MJ/kg (23,300 BTU/lb or 150,000 BTU/gal).

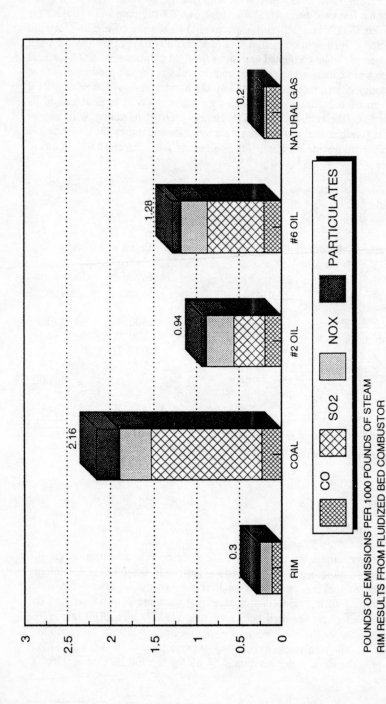

POUNDS OF EMISSIONS PER 1000 POUNDS OF STEAM
RIM RESULTS FROM FLUIDIZED BED COMBUSTOR
DATA FROM NORTH AMERICAN COMBUSTION HANDBOOK

Figure 12. Stack Gas Emissions from Steam Production with Various Fuels (Reproduced with permission from ref. 19. Copyright 1991 Society of Automotive Engineers, Inc.)

Table X. Energy Contents of Automotive Materials

	Energy Content MJ/kg
Steel Sheet (32% Scrap Metal)	44.5
RIM Polyurethane	73.5
Glass Fiber	39.0
Mineral Filler	0.9
Gasoline	54.0
Aluminum (30% Scrap Metal)	240.0

SOURCE: Adapted from ref. 20.

For polyurethane, the energy content of an automotive fascia weighing 4.9 kg (10.8 lbs) and 10% scrap for a total 5.4 kg of material is 397 MJ, Table XI. Recycle of the scrap lowers the energy content of the parts by about 10%, since comparatively little energy is required for the grinding and blending steps. The part energy drops to 360 MJ, a savings of 6.9 MJ/kg, or 37 MJ per part; not a great deal for a single fascia, but significant, if one considers a million parts. Thirty-seven million MJ represents over 680 thousand kilograms of gasoline! The savings are greater when we add in energy recovery.   Part energy requirements drop to 267 MJ, for a total energy saving of 130 MJ, or the equivalence of over 2 kg gasoline energy per part.

Table XI. Part Energy Content of RIM Fascia with Regrind

	RIM	RIM Plus 10% Regrind	RIM with Energy Recovery
Part Weight, kg	4.9	4.9	4.9
Total Weight (10% Scrap), kg	5.4	5.4	5.4
Material Energy, MJ/kg	73.5	66.6	49.4
Part Energy, MJ	397	360	267
Energy Saving, MJ/kg	--	6.9	24.1
Energy Saving/Part, MJ	--	37	130
Energy Saving/1 mm Parts, MJ	--	37*10^6	130*10^6

So far, we have talked about energy savings in the fabrication of RIM polyurethane parts. Let's look now at the energy savings *in use* of these light weight materials, using a body panel - a fender - as an example*(21)*. Again, referring to our materials energy contents (Table X), we have the materials energy for the polyurethane RIM - which contains 25% mineral   filler - at

55.3 MJ/kg, Table XII. (The mineral filler itself has an energy content of only
.9MJ/kg; hence lowering significantly the energy content of the material.)   The
RIM part weighs 3 kg, and the steel part 4.6 kg; and the total part energies,
including scrap are 182.5 MJ for polyurethane and 258 MJ for steel.

Table XII. Materials Energy for Automotive Body Panels

	Recycle RIM	Steel
Raw Material Energy MJ/kg	73.5	44.5
Formulated Material Energy MJ/kg	55.3	44.5
Part Weight, kg	3.0	4.6
Scrap, kg	0.3	1.2
Total Material Used, kg	3.3	5.8
Part Energy, MJ	182.5	258.0

SOURCE: Adapted from ref. 20.

Energy is consumed during the life cycle operation of the automobile.
How much energy depends on several factors, one of which is weight. We have
adopted a conservative value(21) of 4.7 kg of gasoline consumed per kg of
incremental vehicle weight per 163,000 km (100,000 miles); and, since a kg of
gasoline is worth 54 MJ, a kg of incremental weight requires 254 MJ of gasoline
energy.

On this basis, our 4.6 kg steel panel requires 1168 MJ of operational
energy during its life cycle, Table XIII. For the 3 kg polyurethane fender the
total is 782 MJ. The total energy for the steel fender is (part energy + use
energy) 1557 MJ; and for the polyurethane 1007, a difference of 481 MJ
(equivalent to about 9 kg of gasoline) in favor of the polyurethane fender.

Table XIII. Life Cycle Energy Automotive Body Panels

	RIM with Recycle	Steel
Part Energy, MJ	183	258
Use Energy, MJ	762	1,168
Total Energy, MJ	945	1,426
Energy Saved, MJ	481	

SOURCE: Adapted from ref. 20.

Conclusions

To conclude, polyurethanes can be recycled and the recycling saves money and energy resources. The spectrum of available recycle technologies includes the recycle of materials through the use of regrind, compression molding, or extrusion; chemical recycle *via* glycolysis; and thermal recycle with energy recovery. The regrind process offers an attractive route toward recycling RIM manufacturing scrap. It eliminates the cost and ecological problems of landfill, lowers the cost to make RIM parts, and maintains the polymer performance standards. These recycle technologies contribute to overall energy savings and dollar savings in part fabrication, in scrap recycle, in life cycle operation and through energy recovery*(22)*.

Literature Cited

1. Modern Plastics, January 1991, p. 117.
2. Hull and Company,"End Use Market Survey of the Polyurethane Industry in the United States and Canada", prepared for the Society of the Plastics Industry, Inc., Polyurethane Division (May 29, 1990).
3. Miyama, S., *Conservation and Recycling* **1987** *10*, No. 4   pp. 265-272.
4. Baumann, B. D.; Burdick, P. E.; Bye, M. L.; Galla, E. A., SPI-6th International Technical/Marketing Conference; Technomic: Lancaster, PA, 1983, pp. 139-141.
5. Campbell, G. A.; Meluch, W. C., *Environ. Sci. Tech.* **1976** *10*, No. 2, pp. 182-185; Meluch, W. C.; Campbell, G. A., U.S. Patent 3,978,128, (1976).
6. Gerlock, J.; Braslaw, J.; Zimbo, M., Ind. Eng. *Chem. Pro. Des. Dev.* **1984** *23*, pp. 545-552; Braslaw, J.; Gerlock, J., Ind. *Eng. Chem. Pro. Des. Dev.* **1984** *23*, pp. 552-557.
7. Ulrich, H.; Odinak, A.; Tucker, B.; Sayigh, A. A. R. S., *Polym. Eng. Sci.* **1978** *18*, No. 11, pp. 844-848; Ulrich, H.; Tucker, B.; Odinak, A.; Gamache, A. R., *Elast. Plast.* **1979** *11*, p. 208.
8. Bauer, G., RECYCLE 1990, Davos, Switzerland (May 29-31, (1990); Bauer, G., German Patent 2,738,572.
9. Simioni, F.; Bisello, S.; Cambini, M., *Macplas* **1983** *8*, No. 47, p. 42; Simioni, F.; Bisello, S.; Travan, M., *Cell. Polym.* **1983** *2*, pp. 281-283; Simioni, F.; Modesti, M.; Navazzio, G.; *Macplas* **1987** *12*, No. 88, pp. 127-129; Simioni, F.; Modesti, M.; Brambilla, C. A., *Cell. Polym.* **1989** *8*, pp. 387-400.
10. Data from molder PEBRA,GMBH, Esslingen, Germany.
11. Taylor, R. P.; Eiben, R.; Rasshofer, W.; Liman U., International Congress and Exposition, Society of Automotive Engineers, Detroit, MI (February 25 - March 1, 1991), paper 910581.

12. Singh, S. N.; Picolino, E.; Bergenholz J. B.; Smith, R. C., International Congress and Exposition, Society of Automotive Engineers, Detroit, MI, (February 25 - March 1, 1991), paper 910582.

13. Morgan, R. E.; Weaver, J. D., International Congress and Exposition, Society of Automotive Engineers, Detroit, MI (February 25 - March 1, 1991), paper 910580.

14. Wedco, Inc., P.O. Box 145, Grand Junction, TN 38039.

15. Herbold Granulators U.S.A., Inc. Sutton, MA 01590.

16. Pallmann Pulverizer Company, Inc., Clinton, NJ 07012.

17. Salisbury, W. C., U.S. Patent 4,426,348, January 17, 1984.

18. Vanderhider, J. A.; Evans, J. O.; Schrott, B. D., International Congress and Exposition, Society of Automotive Engineers, Detroit, MI (February 24-28, 1986), paper 860513.

19. Myers, J. I.; Farrissey, W. J., International Congress and Exposition, Society of Automotive Engineers, Detroit, MI (February 25 - March 1, 1991), paper 910583.

20. Farrissey, W. J., International Congress and Exposition, Society of Automotive Engineers, Detroit, MI (February 25 - March 1, 1991), paper 910579.

21. Farrissey, W. J.; Morgan, R. E.; Weaver, J., D., SPE 49th ANTEC '91, p. 1522.

22. Gum, Wilson F., *SPI 33rd Annual Polyurethane Technical/Marketing Conference*, Technomic, Lancaster, PA (October 2, 1990).

RECEIVED August 10, 1992

Chapter 23

Technical Aspects of Vinyl Recycling

Bela K. Mikofalvy and H. Khim Boo

Emerging Technologies Department, Avon Lake Technical Center, The
BFGoodrich Company, Avon Lake, OH   44012

Vinyl Uses and Properties

Polyvinyl chloride (PVC) or vinyl, as it is commonly known, is the second
largest volume plastic used world wide: it plays a major role in the plastic
industry and it is also an integral part of plastic recycling opportunities.

In the United States, the yearly consumption of plastics is about 60 billion
pounds and 13% of this total is vinyl. Vinyl has been called the world's most
versatile plastic because of its excellent properties: Its good chemical resistance
combined with its wide compounding flexibility, and ease of fabrication, result
in favorable cost and property performance.

On the one hand, vinyl is a commodity resin; 44% of the vinyl is used for
pipe. On the other, through its compounding versatility, vinyl can be made into
engineered polymers with properties well suited for telecommunications, office
equipment, automotive and other engineered applications. Vinyl is also used
extensively in wire and cable jacketing, house siding and accessories, flooring,
bottles, films and sheets, and other packaging uses - such as blisterpacks and
computer chip tubes. Vinyl products are fabricated by extrusion, calendaring,
molding and coating.

Recycling Rate/Potential

It has been estimated, on the basis of vinyl usage volume in various end-use
applications and on the technical feasiblity of recycling, that there is the
potential to recycle 20% of post-consumer vinyl products.*(1)* This would
yield 1.6 billion pounds of vinyl for recycling. The rate at which the recycled
vinyl volume reaches this potential depends on social and economic factors that
drive the plastic recycling infrastructure which is currently being built, and on
recycled vinyl market demand.

0097–6156/92/0513–0296$06.00/0

To date, vinyl recycling has not developed as rapidly as the recycling of Polyethylene Terephthalate (PET) and High Density Polyethylene (HDPE), but it is a close third. PET and HDPE are to a large extent used in consumer packaging and tend to end up in municipal solid waste after one use.

The total municipal solid waste is 160 million tons a year. This waste contains about 7% plastics by weight based on data collected by Franklin Associates(2) and by others. Vinyl makes up 4% of this 7% plastic portion. This means that 900 million pounds/year or only about 10% of the yearly consumption of vinyl goes into municipal landfills.(3)

The reason for the relatively low level of vinyl in municipal solid wastes is that a significant portion of vinyl goes into longer life semi-durable and durable goods. 65% of vinyl goes into items that have a "use life" of longer than 5 years. Most of the vinyl applications are long term and some remain out of the waste stream for generations. Then, when disposed, they tend to end up in non-municipal solid waste. Recycling of short life products, such as packaging and other disposable goods, has, to date, been the primary focus of recycling but vinyl recycling is rapidly being extended to semi-durable and durable items, with the hierarchy being wire and cable, siding, automobile, then appliances and pipe.

Market demand for vinyl products with recycled content is the most important driving force in the development of more efficient infrastructures, and in the needed technological breakthroughs particularly in sortation. Market demand could significantly increase vinyl recycling volumes. At present vinyl is taking its place among the six most recycled resins and, the vinyl recycling rate is predicted to be in line with the total plastic recycling rate. Specifically, according to a study from the Freedonia Group, Inc.(4) demand for recycled plastics will increase 44%/year through 1994 to 2.3 billion pounds, which is 10 times the growth rate of total plastics demand. Vinyl recycling rate is also predicted to grow at least 10 times the rate of total vinyl growth which is estimated about 3%/year world wide for the period of 1990 - 1995. Plastic News (5) predicts vinyl recycling rates increasing from 0.8% in 1990 to 2% in 1993. If current public interest in plastic recycling continues to grow, there could be within 5 years, 1 billion lbs/year of vinyl for the industry to recycle. Recycling vinyl at this volume will require rapid growth in technological developments during the next few years as well as significant improvements in the infrastructure.

For the purpose of overviewing the state of vinyl recycling, it is useful to categorize recycling into three areas:

1. First, there is the so called "recycling within the process". This refers to re-using the un-utilized vinyl materials, during a run, such as start ups and cut-offs within the same manufacturing operation at the same location.

ASTM - The American Society for Testing and Materials - has allowed this type of recycling, and it has been a well accepted method of operation ever since the start of the vinyl industry. The economics are usually favorable. Today, by most standards, re-using the material within the process is not considered recycling - but "common sense material efficient" operation.

2. The second type of vinyl recycling involves <u>industrial waste</u> which is generated farther down the line in the process and which is exposed to foreign materials, such as labels or printing ink - and, usually requires a degree of cleaning. The recycling rate of the industrial waste stream is about 10% to 25%.

3. The third type of recycling is <u>post-consumer</u>. Recycling plastic waste, which has been used by the consumer, is a relatively new consideration. The infrastructure is currently being put into place by the industry for large commercial scale post-consumer vinyl recycling in the U.S., and it is a fast developing area. The primary reason for slower post-consumer recycling is that sorting vinyl from post-consumer waste is not as simple as picking out milk bottles and soda pop bottles from the waste. All soda pop bottles look alike and are made of PET. Likewise, all milk bottles are made of HDPE and they too look alike. On the other hand, vinyl bottles are more commonly used for a wider variety of products and come in many sizes and shapes. Often vinyl is used interchangeably with other non-vinyl plastics for bottles identical in size, shape and appearance which makes visual separation difficult. Technological developments which are currently under way should soon solve the sortation problem.

Technology

There are three essential elements of recycling:

1. A stable **supply source** which involves collection and sortation.
2. An economically and environmentally sound and reliable **reclaim process**, and
3. Recycled **end-use applications** which yield adequate market values.

All three must be in place for the successful recycling of a material. The technological aspects in each of these areas are being addressed by both industry and academia. The basic vinyl recycling process is shown at the end of this chapter (Figure 1).

Collection and Sortation. At present, the supply source - collection and sortation of post-consumer products - is the weakest link in plastic recycling technology. Most plastics are still hand sorted or, at best, sorted in a semi-automated process which is difficult, uneconomical and low in reliability. The ability to hand separate a vinyl stream from municipal waste was demonstrated

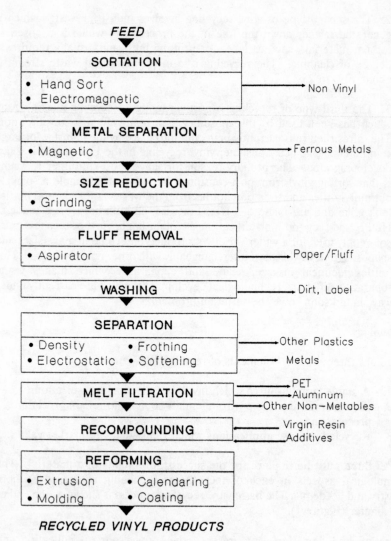

FEED

SORTATION
- Hand Sort
- Electromagnetic

→ Non Vinyl

METAL SEPARATION
- Magnetic

→ Ferrous Metals

SIZE REDUCTION
- Grinding

FLUFF REMOVAL
- Aspirator

→ Paper/Fluff

WASHING

→ Dirt, Label

SEPARATION
- Density   • Frothing
- Electrostatic   • Softening

→ Other Plastics
→ Metals

MELT FILTRATION

→ PET
→ Aluminum
→ Other Non−Meltables

RECOMPOUNDING

← Virgin Resin
← Additives

REFORMING
- Extrusion   • Calendaring
- Molding   • Coating

RECYCLED VINYL PRODUCTS

Figure 1. Vinyl Recycling: The Basic Process

in a pilot study*(6)*. In this pilot study, co-sponsored by wTe Corporation, Dow Chemical Company and BFGoodrich, commingled recyclables in Akron household municipal waste (placed at curb-side) were collected and brought into wTe's Material Recycling Facilities.

Plastics were separated from newspaper, glass, metal, and other residue. Five percent of the recyclables were plastics. The plastics were then further separated manually into the following portions: milk bottles, pop bottles, foam, and vinyl. The remaining plastics were called "mixed". Clear containers which had the SPE code 3 or had a "smile" at the parison pinch off, which occurs during extrusion blow molding, were put into the vinyl stream. The vinyl thus collected amounted to 1% of the total plastics. This method resulted in 80% accuracy in identifying vinyl, as the "vinyl stream" contained almost 10% PET plus a number of other materials totaling another 10%.

Milk and pop bottles, some vinyl bottles, and polystyrene items could easily be identified, but 31% of the total plastic containers could not be identified by hand sorting and were lumped together as the mixed portion. When analyzed by infrared, the mixed portion contained 10% vinyl. The 1% separated vinyl therefore, represented only 25% of the total vinyl collected at the curbside, while 75% of it ended up in the mixed portion.

Because hand separation is not effective enough, improved sortation technologies, based on the differences of various inherent properties of different plastics are rapidly being developed. There are six technologies in various stages of development:

1.   Density,
2.   Electromagnetic,
3.   Selective Dissolution,
4.   Electrostatic,
5.   Frothation Separation, and
6.   Softening Point Separation.

The electromagnetic based processes are geared to sorting whole items as collected; the other methods are focused on sorting the flakes after granulation.

The **density separation** method is based on the differences in specific gravities of various plastics ranging from 0.96 to 1.37. Simple float or sink baths or cyclone type density separation do a good job in separating PVC from all the plastics except PET. Because the specific gravities of PVC and PET overlap, these two polymers cannot be separated by density. Typically, PVC density ranges between 1.30 to 1.37 and PET's density range is 1.34 to 1.36.

Recent developments using supercritical liquids*(7)* promise to broaden the range of density-based separation processes by allowing fluid density control to within +/- 0.001 g/cm^3 which would make it possible to sort PVC from PET by density except in cases (i.e., PET copolymers) where the densities of these two polymers overlap.

Electromagnetic separation is based on identifying the chlorine atom in PVC by x-ray. An example of an electromagnetic separator is the Asoma Vinyl Bottle Sorter Model 652-D*(8)*. This detector is based on x-ray fluorescence of chlorine. A simplified principle of this method is that an x-ray is used to knock an electron out of the chlorine atom orbital in the PVC molecule. When an electron returns to the chlorine orbital, it releases a unique amount of energy as an x-ray of wavelength unique to chlorine. The detector measures the amount of these chlorine x-rays. Except for polyvinylidene chloride that may be used as a barrier layer in some packaging and some flame retardant plastics (usually non-packaging) the chlorine x-ray is unique to vinyl.*(9)*

The National Recovery Technologies Process*(10)* which is currently being commercialized and the Govoni process*(11)* in Italy are also based on identification by x-ray beams. These processes are geared to sorting items prior to granulation or flaking. When the detector identifies vinyl it triggers a device which kicks off the vinyl bottle from the conveyor.

Selective dissolution is being developed at Rensselaer Polytechnic Institute.*(12)* In a given solvent, such as Xylene, different polymers dissolve at different temperatures. As the solvent temperature is increased, the various polymers dissolve in sequence and can be removed from the solvent. The advantage of the solvent technique is that it cleans, separates, and reprocesses the resins to their original virgin states - all in one process. Our studies showed that PVC recycled by this process has properties comparable to virgin resin. The disadvantage is the anticipated high cost of commercial implementation to ensure an environmentally sound operation with solvents.

The **electrostatic** approach is based on the fact that the various plastic types behave differently when subjected to an electrostatic charge. Examples are the DevTech*(13)* process and Professor Inculet's*(14)* work at the University of Western Ontario, Canada.

Sorting vinyl from a mixture of flakes by various forms of **frothation separation** are also being developed. This technology can be employed to separate two types of plastics, with similar densities, such as PVC and PET. It usually requires the addition of a surfactant or some other additive to modify the surface properties of the plastic granulates prior to frothation. The Goodyear Company's*(15)* PVC separation process uses N_2 or air draft after conditioning the flakes. Recovery's *(16)* so called main stage process, removes ground PVC also in a frothation system.

Differences in the **softening points** of various polymers provides another separation method with potential to separate unplasticized PVC, which has a melting point of around 200°C, from PET with a melting point of 260°C. An example of softening point separation is the recently patented process by Refakt(17) in Germany, which can separate small amounts of PVC from PET.

Reclaim Process.

Vinyl Stream Purification. After sorting the collected items by either manual or automatic separation, the vinyl stream is ground into flakes. The reclaiming process consists of purification of the vinyl flakes by washing, usually in conjunction, with a density based separation. A good example of a vinyl stream purification is the use of a calcium nitrate density bath(18). In this process, vinyl bottles from municipal solid waste yielded vinyl with <2% non-vinyl contamination. The vinyl was granulated to ½ inch flakes, then aspirated to remove any fines, thin films, and paper.

The granulated material was then washed with agitation in 80°C water which had a 1% concentration of detergent. The floaters and lower specific gravity materials were then flushed out with the paper using clear water. The sinkers (specific gravity > 1.00) were screened, dewatered and placed into a calcium nitrate bath.

A calcium nitrate solution at 1.35 specific gravity was used to sink the aluminum and some PET and float the vinyl at less than 1.35 specific gravity. Floaters were then put through a calcium nitrate solution at 1.30 specific gravity to float out the PC and some PET copolymer. The sinkers - purified vinyl - were recovered, washed and dried.

The resulting purified vinyl showed excellent tensile strength, elongation, modulus, heat deflection temperature and thermal stability with properties suitable for drain, waste and vent (DWV) pipes.

If further purification is desired melt filtration can be employed.

Melt Filtration. One of the difficulties in plastic recycling is the purification of the recycled materials to near zero contamination levels. After flake purification, the vinyl typically contains a few tenths of a percent contaminants. While for some applications, the presence of <1% contaminants may not be critical, other applications require <0.1% contaminants. This is especially true for applications with thin walled parts.

Melt filtration is a filtration process that takes place at the polymer melt stage to remove contaminants. A screen is placed just before the die to screen out materials that are not meltable at vinyl's processing conditions. This technology is particularly useful for screening out metals, papers, fibers and incompatible polymers of higher melting temperature than vinyl. Melt filtration has been used in the plastic industries for virgin compounds to improve compound quality by screening out gels or undispersed agglomerates. To some extent, it has also been used in the extrusion step, primarily on flexible vinyl, to remove contamination.

Melt filtering a virgin vinyl compound is much simpler than melt filtering recycled vinyl because of the large difference in the contamination level of these two materials. While virgin vinyl may contain impurities in the thousandth or hundredth of a percent and thereby would not contribute to any pressure build-up during a reasonable processing time frame, the contaminants in a recycled vinyl can plug the screen in minutes. The recycled PVC stream needs to be purified to < 1% non-meltables contamination before melt filtration can be effectively performed.

Melt filtration is a relatively standard practice in the recovery of recycled HDPE and PET. HDPE and PET have sharp melting points, relatively low melt viscosities and excellent thermal stability at melt temperatures. This permits good melt filtration both with stationary and continuous screening systems with minimal impact on extrusion rates and back pressure at the extruder die. PVC is an amorphous polymer which melts and flows over a range of temperatures. The melt viscosities of PVC tend to be somewhat higher. In addition, it's degradation temperature is relatively close to it's processing temperatures. Therefore, it is critical that surfaces in contact with the PVC melt be streamlined to minimize stagnation areas where polymer degradation can occur.

Contamination level, screen pack configuration, screw speed and temperature effect on the melt filtering process and on the recycled vinyl quality are being studied(19). One of the important processing factors in melt filtration is the pressure build up at the screen. The pressure build up is directly related to the contamination level.

In principle, the finer the screen used for melt filtration, the smaller the size of contaminants it can screen out. Thus, the quality of the melt filtered material is primarily dependent upon the size of the screen used. The effectiveness of the melt filtration is greatly reduced with recycled vinyl which contains crosslinked rubbery materials, such as the vinyl recycled from wire and cable. Elastic particles deform themselves under high pressure and go through screen meshes smaller than their stationary particle size.

Nevertheless, melt filtering of recycled PVC, whether it is rigid or flexible, is possible. Flexible PVC is easier to melt filter because of lower viscosity and better thermal stability. A continuous screen changer with a design that prevents dead spots is preferred for PVC. In light of recent needs for recycled vinyl, several companies, such as HiTech, Key Filters, Beringer, Patt Filters and Gneuss, are offering continuous melt filter systems for both flexible and rigid PVC. Through a simultaneous recompounding and melt filtering process, cleaning PVC to a contamination-free stage is feasible. The economics of the process is primarily dependent on the screen cost.

An alternate approach to melt filtering is grinding or pulverizing the contaminants to make them small enough to act as fillers. Grinding to sub mm size is expensive, and in some cases, physical properties are still poor even after fine grinding the non-meltable contaminants to <0.1mm*(20)*.

Re-Compounding. There are both processing and quality considerations in re-compounding recycled vinyl. The three important processing considerations are:

1.   Aspect ratio of the recycled vinyl granulate,
2.   Additive take-up, and
3.   % moisture content.

An aspect ratio, or the ratio of length to thickness of the granulated particles that is too high causes bridging in the extruder hopper. For example, when credit cards are ground to 3/8" to 1/2" granulates, the aspect ratio is about 4 to 10. When mixed with PVC powder, or typical PVC pellets and/or cubes, the ground up credit cards cause bridging and prevent an even flow to the extruder. Microatomization to about 7 mesh (0.06") or 1.5mm to reduce aspect ratio to 0.5 to 1 is required to eliminate bridging.

In the re-compounding operations, adding liquid stabilizers and plasticizers to recycled PVC can present problems since the recycled PVC has already been fused and lacks porosity. The problem of additive take up can be minimized by changing compounding procedures and/or conditions such as pre-heating the recycled PVC prior to adding the liquid. Also, a high porosity virgin PVC powder resin can effectively be used as a carrier for the liquid additive.

Generally, moisture content of PVC compound should be <0.2% and even lower for molding to prevent surface defects. The recycled vinyl flakes should be dry enough not to increase the total moisture content in the compound above the acceptable level. Typically, <0.5% moisture in the recycled flakes is acceptable at 25% recycled content.

In deciding on end-use applications for a given recycled vinyl material, the first step is to establish the stability of the recycled product. This is the most important quality consideration. A good way to determine the heat stability is the Dynamic Thermo Stability Test*(21)*.   Using this test, if burn (severe discoloration) is observed in less than 12 -15 minutes on rigid PVC or less than 30 - 60 minutes on flexible PVC, stabilizer should be added to the recycled material.

The second important step is to characterize the properties of the recycled material. The important properties are: tensile strength, elongation, modulus, hardness, impact strength, heat deflection temperature (HDT), and viscosity (melt index). Elongation is very sensitive to even low levels of contamination. Modulus, HDT, and hardness are good indicators of plasticizer levels. Impact is sensitive to impurities. Melt index indicates if the material can be used for injection molding.

The third step is the actual recompounding to make the recycled material fit the intended "new" end-use application. Properties that can easily be adjusted are:

Impact Strength	-	By adding impact modifier
Stability	-	By adding stabilizer
Melt Index	-	By blending with different molecular weight resins, or
	-	By adding plasticizer
Hardness	-	By adding plasticizer to decrease
	-	By adding unplasticized (rigid) vinyl to increase

Recycled Vinyl Products

Vinyl from both industrial waste and post-consumer products can be recycled into good quality recycled products.

A 1990 analysis*(22)* of potential end-use markets for recycled PVC shows the existence of over 90 vinyl product lines, many of which could use recycled material. Products identified in the analysis would consume nearly half a billion lbs/year of recycled vinyl.

The feasibility of manufacturing a variety of recycled products from a properly cleaned vinyl stream from municipal solid waste has been demonstrated*(23)*. For example, bottles have been recycled into bottles, drainage pipe, and drainage pipe fittings, with good appearance and properties. The physical properties of the recycled bottle compound are quite comparable to the virgin compound.

Pipe compounds made with recycled bottle content generally have properties suitable for some pipe applications. Tensile strength and modulus tends to decrease with increased recycled content, but at levels up to 30% recycled content, the compounds meet the standard. ASTM has groups working to develop definitions and a standard for use of recycled vinyl in drain pipe.

Extrusion properties indicate that the recycled bottle compounds fuse faster than virgin pipe compound, causing higher screw amperage. The compound therefore, should have increased lubrication or the extruder temperature should be decreased.

In general, for molded applications, non-meltable particles less than 0.1mm in size do not present an appearance problem because the mold gives a smooth surface. In extrusion, even small non-meltable particles cause streaking. However, even if the surface is smooth, the effects of non-meltable particles on physical properties must be assessed. Particles which do not adhere well to vinyl such as PET or crosslinked PE, will reduce physical properties while particles which have surface adherence to vinyl, i.e., crosslinked PVC will have minimal effect.

Recycling of vinyl wire and cable into a variety of applications, including traffic cones, drainage pipe and back into wire and cable, has also been demonstrated with promising results for commercial applicability(24). One school of thought (25) is "that the recycling cost structure, the product properties, the advent of recycle-content legislation and the price of bottle compound dictate that recycled PVC bottles be returned to the packaging stream". Others believe that the primary direction should be to convert disposables to durables(26).

Effects of Impurities

A molding study(27) involving a molded vinyl article which was reground then re-molded six times showed that recycling rigid vinyl without any impurities tends to improve two of the critical mechanical properties, i.e., impact strength, and tensile elongation. At each reprocessing step key properties were measured and compared with the properties of moldings from virgin materials. Impact strength and tensile elongation, became measurably superior at the first reprocessing step, improved still further on the second and third steps and maintained this substantially improved performance throughout the six reprocessing cycles. The same pattern was followed by the mold flow, or melt viscosity. Mechanical properties known to be relatively unaffected by degradation, i.e., tensile strength and heat deflection temperature, were virtually unaffected. As expected, processing stability, as measured by dynamic thermal stability and yellowness, deteriorated by approximately the same amount at every reprocessing step. At no time did the compound become difficult to mold at normal injection molding machine settings.

Impurities, however, tend to decrease the physical properties. Most noticeable is the decrease in elongation. The amount and type of impurities which can be tolerated needs to be determined for each recycled application and, in addition to appearance, should be based on property performance.

A number of studies have been published on the effects of impurities on the properties of recycled vinyl. One study *(28)* showed that simple removal of paper and oil by washing will allow the vinyl bottle granulate to recover most of the properties of the original bottle compound. Paper and oil impurities most significantly effect the Izod Impact of the recycled bottles. Another study*(29)* concluded that all contaminants must be removed from PVC powder recovered from waste bottles to give good quality pipe.

The physical properties of recycled vinyl packaging material from municipal solid waste at various stages of purification showed*(30)* that the purified vinyl has good properties and could be considered for use in non-pressure pipe. It also meets the mechanical properties required for sewer fittings.

Since it is primarily a bottle compound, recycled compound from municipal waste, even though it is much tougher, would not be expected (at 100% use level) to meet all the ASTM requirements specified for drain, waste, and vent pipe. It is well understood that adding or increasing the level of a rubbery impact modifier, such as is present in bottles, to a thermoplastic resin will increase its toughness as well as decrease its tensile strength and decrease its modulus. Impact modifiers also tend to cause more swelling in sulfuric acid and oil. The bottle compounds and the recycled vinyl bottle are more impact modified than that specified in the ASTM standard for drain, waste, and vent or in the ASTM standard on sewer pipe.

A comparison of properties of recycled vinyl from municipal solid waste after grinding, washing and specific gravity separation (contamination level of < 1%), to virgin bottle compound shows that except for lower tensile elongation and oven thermal stability, the physical properties are retained. Any unmeltable impurities are usually small enough to be unnoticed and do not cause holes or other physical defects. For example, PET particles, unless ground into <0.1mm, cause holes in an extruded product. In general, the tensile strength decreases with increasing PET contact. The rate of decrease depends on the particle size of PET. At 0.1mm these effects are minimum. PET contamination above 1% however, causes severe drop in Izod Impact and Elongation even if PET is finely ground.*(31)*

The future of vinyl recycling is being built. Collection infrastructures are being installed. Sortation technology is being developed and proven. Proper sortation can insure a steady supply source for recycled vinyl. Technical and infrastructure developments in the reclaim process have proven the ability to clean, to purify, and to supply a quality recycled vinyl. Several recycled vinyl end-use applications have already been proven feasible with a growing acceptance and demand. With these developments, the future of vinyl recycling is bright.

References

1. Lantos, P., The Target Group, personal communication, 1990.
2. Franklin Associates, Ltd. and The Garbage Project; Estimates of the Volume of Municipal Solid Waste and Slected Components (Council For Solid Waste Solutions) 1989.
3. Mikofalvy, B.K. and Summers, J.W., The B.F.Goodrich Chemical Company; Recycling Vinyl Applications and Preliminary Economics USA [Speech] (Recycle '90; Davos, Switzerland, 1990).
4. Freedonia Group, Inc., Plastic Recycling to 1994.
5. Charnas, D.; "PET,HDPE Lead the Pack in Recycled Resin Use," Plastic News, Sept. 17, 1990. p.13.
6. Summers, J.W.; Mikofalvy, B.K.; Wooton, G.V.; Sell, W.A. J. of Vinyl Technol.6., 1990, Vol. 12, No. 3. pp 154-160.
7. Super, M.S. et al, University of Pittsburgh, Separation of Thermoplastics Using Nitrogen and Supercritical Co_2 as a Precursor to Recycling. [Speech] (Plastics Recycling Division: Society of Plastics Engineers, Inc., 1991), pp 11-14.
8. Asoma Instruments, Inc., Asoma Instruments Vinyl Bottle Sorter Model 652-D [Flier] (Austin, Texas: Asoma Instruments, Inc. 1990).
9. Summers, J.W.; Mikofalvy, B.K; Little, S. J. Vinyl Technol., 1990, Vol 12, No. 3.
10. National Recovery Technologies, Inc., Sorting It Out (Nashville, TN: National Recovery Technologies, Inc., 1991)
11. Govoni SpA, Technology for the Plastic Industry, (Italy: Govoni SpA, 1989)
12. Lynch, J.C. and Nauman, B. E. Separation of Commingled Plastics By Selective Dissolution. [Speech] (Society of Plastics Engineers, Inc., 1989).
13. DevTech Labs, Inc., Amherst, New Hampshire.
14. Prof. Inculot, Director, Applied Electrostatics Research Centre, Faculty of Engineering Science, The University of Western Ontario, London, Ontario, Canada.
15. Goodyear Tire and Rubber Co., Akron, Ohio.
16. Recovery Processes Inc., Park City, Utah.
17. REFAKT, Meckesheim, Germany.
18. Summers, J.W.; Mikofalvy, B.K.; Wooton, G.V.; Sell, W.A., J. of Vinyl Technol. 1990 Vol 12, No. 3.
19. Boo, H.K., Mikofalvy, B.K.; Mittendorf, D.H.; Sell, W.A.; The B.F.Goodrich Company, Melt Filtration of Recycled PVC Paper to be presented at ANTEC, 1992. Society of Plastics Engineers, Inc.
20. Summers, J.W.; Mikofalvy, B.K.; Krogstie, J.M.; Sell, W.A.; Separating Vinyl Packaging From Municipal Solid Wastes; [Speech] (Society of Plastics Engineers, Regional Technical Meeting, Palisades, NJ Section, Oct. 10, 1990).
21. The B.F.Goodrich Company; Cleveland, Ohio: In-house Dynamic Thermal Stability Test.

22.   Bennett, R.A.; Lefcheck, D.L.; End Use Markets for Recycled Polyvinyl Chloride [Report] Prepared by The College of Engineering, The University of Toledo - Toledo, Ohio July 1990.

23.   Summers, J.W.; Mikofalvy, B.K.; Boo, H.K.; Krogstie, J.M.; Sell, W.A.; Rodriguez, J.C.; The B.F.Goodrich Company *Examples of Recycled Vinyl Products*: Paper to be presented at ANTEC, 1992. Society of Plastics Engineers, Inc.

24.   Mikofalvy, B.K.; Boo, H.K.; Zemanek, M.J.; Krogstie, J.M.; Sell, W.A.; Mittendorf, D.J.; *Traffic Cones from Recyled Material* Paper to be presented at ANTEC, 1992. Society of Plastics Engineers, Inc.

25.   Carroll, W.F., Jr: *J. Vinyl Technol.*, **1991**, Vol 13., No. 2, p. 96.

26.   Krause, F.E.; The B.F.Goodrich Company, BFGoodrich Recycling Strategy: Disposables to Durables; [Paper] (Recyclingplas VI Conference; Washington, D.C., 1991).

27.   Lloyd, P., The B.F.Goodrich Company, personal communication, 1991.)

28.   Pazur, A.S.; *J. Vinyl Technol.* 1988, Vol 11.

29.   Burgaud, P.; *J. Appl. Polymer Sci.*

30.   Summers, J.W.; Mikofalvy, B.K.; Wooton, G.V.; Sell, W.A.; *J. Vinyl Technol.* **1990**, Vol 12, No. 3.

31.   Summers, J.W.; Mikofalvy, B.M.; Wooton, G.V.; Sell, W.A.; *J. Vinyl Technol.* **1990**, Vol. 12, No.3.

RECEIVED August 27, 1992

INDEXES

Author Index

Affiliation Index

Subject Index

Production: Bruce Hawkins
Indexing: Deborah H. Steiner
Acquisition: Anne Wilson
Cover design: Sue Schafer

Printed and bound by Maple Press, York, PA